AF707530

AUGUSTUS DE MORGAN AND THE LOGIC OF RELATIONS

The New Synthese Historical Library
Texts and Studies in the History of Philosophy

VOLUME 38

Series Editor:

NORMAN KRETZMANN, *Cornell University*

Associate Editors:

DANIEL ELLIOT GARBER, *University of Chicago*
SIMO KNUUTTILA, *University of Helsinki*
RICHARD SORABJI, *University of London*

Editorial Consultants:

JAN A. AERTSEN, *Free University, Amsterdam*
ROGER ARIEW, *Virginia Polytechnic Institute*
E. JENNIFER ASHWORTH, *University of Waterloo*
MICHAEL AYERS, *Wadham College, Oxford*
GAIL FINE, *Cornell University*
R. J. HANKINSON, *University of Texas*
JAAKKO HINTIKKA, *Boston University, Finnish Academy*
PAUL HOFFMAN, *Massachusetts Institute of Technology*
DAVID KONSTAN, *Brown University*
RICHARD H. KRAUT, *University of Illinois, Chicago*
ALAIN DE LIBERA, *École Pratique des Hautes Études, Sorbonne*
DAVID FATE NORTON, *McGill University*
LUCA OBERTELLO, *Università degli Studi di Genova*
ELEONORE STUMP, *Virginia Polytechnic Institute*
ALLEN WOOD, *Cornell University*

The titles published in this series are listed at the end of this volume.

DANIEL D. MERRILL

Department of Philosophy, Oberlin College, USA

AUGUSTUS DE MORGAN AND THE LOGIC OF RELATIONS

KLUWER ACADEMIC PUBLISHERS
DORDRECHT / BOSTON / LONDON

Library of Congress Cataloging-in-Publication Data

Merrill, Daniel D. (Daniel Davy)
Augustus De Morgan and the logic of relations / Daniel D. Merril.
p. cm. -- (The New synthese historical library ; 38)
ISBN 0-7923-0758-5 (alk. paper)
1. De Morgan, Augustus, 1806-1871--Contributions in logic of relations. 2. Inference. 3. Syllogism. 4. Logic, Symbolic and mathematical--History--19th century. I. Title. II. Series.
BC185.M47 1990
160'.92--dc20 90-34935

ISBN 0-7923-0758-5

Published by Kluwer Academic Publishers,
P.O. Box 17, 3300 AA Dordrecht, The Netherlands.

Kluwer Academic Publishers incorporates
the publishing programmes of
D. Reidel, Martinus Nijhoff, Dr W. Junk and MTP Press.

Sold and distributed in the U.S.A. and Canada
by Kluwer Academic Publishers,
101 Philip Drive, Norwell, MA 02061, U.S.A.

In all other countries, sold and distributed
by Kluwer Academic Publishers Group,
P.O. Box 322, 3300 AH Dordrecht, The Netherlands.

Printed on acid-free paper

All Rights Reserved
© 1990 Kluwer Academic Publishers
No part of the material protected by this copyright notice may be reproduced or utilized in any form or by any means, electronic or mechanical including photocopying, recording or by any information storage and retrieval system, without written permission from the copyright owner.

Printed in the Netherlands

TABLE OF CONTENTS

PREFACE vii

CHAPTER I. The Traditional Syllogism 1
1. Whately and the Revival of Formal Logic 2
2. Euclid and the Syllogism 10
3. Reid, Hamilton and Mansel on Relational Inferences 15

CHAPTER II. First Thoughts on the Copula 26
1. The Two Copulas 26
2. First Notions of Logic 35
3. Relations and Identity 43

CHAPTER III. Generalizing the Copula 48
1. The Abstract Copula 49
2. The Bicopular Syllogism and the Composition of Relations 60
3. Oblique Inferences and De Morgan's *Dictum* 79

CHAPTER IV. The Problem of Form and Matter 89
1. "Sundry Perversions of the Syllogistic Form" 90
2. The Material Copula 95
3. De Morgan's Response 99
4. The Issues 103
5. Heads and Tails 110

CHAPTER V. The Logic of Relations 113
1. Philosophical Preliminaries 114
2. General Logic of Relations 117
3. Properties of Relations 124
4. Singular Relational Syllogisms (Unit Syllogisms) 129
5. Quantified Relational Syllogisms 136
6. The Limited Unit Syllogism 143
7. The Ordinary Syllogism and the Relational Syllogism 145

CHAPTER VI. The Logic of Relations and the Theory of the Syllogism 149
1. The Two Views 150
2. Objective View–The Basic Account 151
3. Objective View–The Relational Form 156
4. The Subjective View 164

CHAPTER VII. Logic and Mathematics 170
1. "A Mathematical Logic" 170
2. Algebraic Techniques and Analogies in Logic 174
3. Logic and Geometrical Proof 176
4. Logic and Algebraic Reasoning 180
5. Form in Algebra and Logic 183
6. Conclusions 193

CHAPTER VIII. A Rigorous Formulation 196
1. Basic Issues 196
2. The System D 200
3. Properties of Inclusion and Identity 202
4. De Morgan's Basic Identities 206
5. Theorem K 208
6. Properties of Relations 212
7. Additional Inclusion Laws 217
8. The Full System of Three-Relation Terms 219
9. De Morgan's Logic with Identity 221
10. More Properties of Relations 223
11. A Surrogate for Quantification Theory 229
12. Postscript–1864 235
13. De Morgan's Conjectures 240

NOTES 245

INDEX 255

PREFACE

The middle years of the nineteenth century saw two crucial developments in the history of modern logic: George Boole's algebraic treatment of logic and Augustus De Morgan's formulation of the logic of relations. The former episode has been studied extensively; the latter, hardly at all. This is a pity, for the most central feature of modern logic may well be its ability to handle relational inferences. De Morgan was the first person to work out an extensive logic of relations, and the purpose of this book is to study this attempt in detail.

Augustus De Morgan (1806–1871) was a British mathematician and logician who was Professor of Mathematics at the University of London (now, University College) from 1828 to 1866. A prolific but not highly original mathematician, De Morgan devoted much of his energies to the rather different field of logic. In his *Formal Logic* (1847) and a series of papers "On the Syllogism" (1846–1862), he attempted with great ingenuity to reformulate and extend the traditional syllogism and to systematize modes of reasoning that lie outside its boundaries. Chief among these is the logic of relations. De Morgan's interest in relations culminated in his important memoir, "On the Syllogism: IV and on the Logic of Relations," read in 1860.

De Morgan made important contributions to the study of the categorical syllogism, and he invented many novel notations and rules for this purpose. The sheer variety of his approaches to the syllogism give his work an almost chaotic quality, despite the systematic motivations behind it. In the course of these investigations, he introduced into logic the concept of the universe of discourse and rediscovered what have come to be called De Morgan's laws–e.g., that the complement of the intersection of two classes is the union of the complements of the two classes.

De Morgan's greatest achievement in logic was undoubtedly his discovery of the logic of relations. Some earlier logicians, including Aristotle, had noted the existence of relational arguments, such as, "Every circle is a figure; therefore, anyone who draws a circle draws a figure." It appears, though, that De Morgan was the first logician to sense the pervasiveness of relational arguments and to see their systematic importance for logic.

This belief in the importance of relational arguments led De Morgan to create much of the logic of relations as we know it today. He made the fundamental discovery that relations could be compounded so that, for instance, one can compound the relations *brother* and *parent* to get their relative product, brother of a parent–i.e., *uncle*. Once this and other forms of relational composition are introduced, a vast array of inferences comes into view. Among them is the inference which De Morgan called Theorem K, which states that if the product of a relation L with a relation M is included in a relation N, then the product of the complement of N with the converse of M is included in the complement of L. This means that the statement "Every brother of a parent of a person is an uncle of that person," implies "Every non-uncle of a child of a person is a non-brother of that person."

De Morgan was also the first person to study the properties of relations, such as transitivity and symmetry (which he called convertibility). Once again, complex logical inferences abound. De Morgan notes, for instance, that the transitivity of the ancestor relation implies that "Among non-descendents are contained all ancestors of non-descendents."

Such inferences are hardly trivial, and De Morgan's memoir on the logic of relations is filled with similar examples. We must admire not only his ingenuity in constructing the logic of relations but also his ear for valid inferences.

The logic of relations has long since been incorporated into quantification theory, but it remains of interest in its own right. For De Morgan, it was as if he had discovered a new logical world. After developing this new logic, he proclaimed, "And thus in logic, as in mathematics, the horizon opens with the height gained: generalization suggests detail, which again suggests generalization, and so on *ad infinitum*"(S4,235).

While realizing the great importance of the logic of relations for later developments in logic, it is important to place De Morgan's pioneering work in the context of his own thought. This is one of the main objectives of this book. We will see how De Morgan's logic of relations grew out of two idiosyncratic features of his thought: an analysis of propositions which turned all categorical propositions into relational propositions; and the resulting attempt to work out syllogistic logic within the framework of the logic of relations. This framework both motivated and limited De Morgan's logic of relations.

Chapter One sets the stage. I first consider some central features of Archbishop Whately's *Elements of Logic* (1826), which is often credited with restoring the study of formal logic in Great Britain. Also included is a discussion of the importance of relational inferences for the logic of geometry and of attempts by Reid, Hamilton and others to deal with such inferences.

In Chapters Two through Four, I trace the development of De Morgan's very distinctive relational account of judgment and inference. He moves from what I will call *the doctrine of the two copulas* to the *doctrine of the abstract copula* to the *bicopular syllogism and the composition of relations*. In the process, all judgments, even subject-predicate judgments, are seen as having a relational form. The traditional copula comes to express merely another material relation, no different from, say, "is joined to." As a result, all inferences, even those of the traditional syllogism, are grounded in the logic of relations. De Morgan's radical views on form and matter in logic become clarified only in replying to the criticism that he had gone beyond formal logic.

This relational analysis of inference cried out for a logic of relations, which is the subject-matter of De Morgan's 1860 paper. In Chapter Five I outline the main features of this logic, and in Chapter Six I show how he applied it to several analyses of the traditional syllogism. All of this is described in considerable detail, so that the reader can get an idea of how the system is supposed to work. It is also evaluated critically.

De Morgan did not develop this logic systematically. In Chapter Eight, I formulate a rigorous version of this logic, providing a full list of assumptions and a logical order for the theorems. Additional theorems are deduced, and the system is augmented in various ways, so that even a surrogate for quantification theory is obtained.

It is natural to wonder about the extent to which De Morgan's work as a mathematician was linked to his logic of relations. In Chapter Seven I examine this issue from several points of view. While De Morgan's concern for the nature of mathematical reasoning had only a marginal impact on his logic of relations, his relational approach to judgment and inference could well have been suggested by very general mathematical considerations. But there is no evidence that the 1860 logic of relations was substantially influenced by his study of mathematics.

This book grows out of my earlier studies in the history of the logic of relations which were supported by Oberlin College, through sabbatical leaves and a Research Status appointment. I greatly appreciate a fellowship which I received from the American Council of Learned Societies (supported, in turn, by the National Endowment for the Humanities), which made it possible for me to spend the 1981–82 academic year in London examining De Morgan's manuscripts and letters. Most of these papers are housed in the Paleography Room in the Senate House Library of the University of London. I wish to express my deep appreciation to the staff members of the Paleography Room, who were unfailingly courteous and helpful in responding to my innumerable requests for assistance during the year in London.

Many other libraries provided access to other portions of De Morgan's correspondence, and I acknowledge their help as well. These include: University College of the University of London; the British Library; the University Library and University Archives of Cambridge University; St. John's College and Trinity College, Cambridge University; The Bodleian Library, University of Oxford; Trinity College, Dublin; The Royal Society of London; and The Royal Astronomical Society. I am also in the debt of the Oberlin College Library, both for its fine collection of works on 19th century logic and for the assistance of its interlibrary loan staff.

Portions of this book have been presented at departmental seminars at the London School of Economics, Chelsea College of the University of London, and the State University of New York at Buffalo; and, at the conference on "The Birth of Mathematical Logic," held in March, 1983, at the State University of New York College at Fredonia. This book has benefited from the discussion which followed each of these presentations.

I am indebted to Peter Heath for his excellent edition of De Morgan's logical writings and for his encouragement in the initial stage of this project. It has been a pleasure to discuss topics from this book with John Corcoran, Randall Dipert, Ivor Grattan-Guinness, Calvin Jongsma, and Joan Richards. In faculty seminars at Oberlin College, I have been helped by comments from my colleagues Norman Care, Robert Grimm, David Love, Alfred MacKay, Peter McInerney, and Ira Steinberg. I am especially indebted to John Corcoran for his detailed comments on an earlier draft of this book. Any errors which remain are, of course, my responsibility.

I wish to thank the Open Court Publishing Co. for allowing the quotations from their reprint editions of De Morgan's *Formal Logic*, edited by A.E. Taylor, and *On the Study and Difficulty of Mathematics*. The quotations from *On the Syllogism and Other Logical Writings*, edited by Peter Heath, appear with the permission of Routledge.

Many thanks go to Karen Barnes and Linda Robinson, and to Candace Shaw and Marian Zelman, for the gracious and efficient way in which they did the word processing needed for the several drafts of this rather complex manuscript.

Finally, I am deeply grateful to my wife, Marly, for the support which she has given me over the years. In a project which took considerably longer than expected, her understanding and encouragement have been invaluable.

THE TRADITIONAL SYLLOGISM

This chapter provides some background for understanding the two most significant innovations in De Morgan's logic of relations. The first and most obvious point is De Morgan's recognition of irreducibly relational inferences: without relational inferences that cannot be handled adequately by traditional logic, there would be little point in devising a new logic for that purpose. His other major claim concerns the very subject matter of logic. Traditional logic is often said to be *formal* logic; and, De Morgan's major logic book even carries just this title. Yet, De Morgan's conception of formal logic is very novel, if not idiosyncratic, and this novelty provides the conceptual underpinning for his logic of relations. In his view, the logic of relations provides the structure for even the inferences of traditional logic.

To set De Morgan's innovations in context, we shall deal with two different topics. The first of these is Richard Whately's much-used *Elements of Logic*, which was first published in 1826. It is important for our purposes because its conservatism contrasts vividly with De Morgan's approach to logic in two important ways. In the first place, it provides a rigorous characterization of the traditional notion of logical form. This ties the logical form of a proposition very closely to its verbal expression and especially to its quantifier and copula. As a result, the validity of an argument will depend solely upon the logical form of the propositions which make it up–and will, in fact, follow from their "mere force of expression." By generalizing the copula and giving to all propositions a relational logical form, De Morgan will propose a radically different version of the traditional distinction between the formal and the material components of a proposition.

Whately is conservative in a second way as well. He limits the logical forms of propositions to those recognized in traditional syllogistic logic, and he claims that the traditional theory of the syllogism suffices for all of formal logic. De Morgan's logic of relations is motivated by the belief that this is not adequate and that even the logic of the categorical syllogism can be properly understood only by embedding it in the logic of relations.

The second topic of this chapter is one challenge to the adequacy of the traditional theory of the syllogism. If Whately is right in saying that this theory can handle all valid deductive arguments, it ought to be adequate for Euclidean geometry, the paradigm of "demonstrative science." In section two we will examine one historic attempt to "syllogize" Euclid. Any such attempt will present many problems, but the most significant problem for our purposes is that of syllogistically representing one type of relational inference: that which embodies the transitivity of relations such as *is equal to* and *is greater than*. In the final section, we will note that Thomas Reid was well aware of this problem, and that he went so far as to say that relational propositions do not have the same logical form as subject-predicate propositions. We will also examine the attempts by Sir William Hamilton and Henry L. Mansel to provide subject-predicate analyses of relational propositions and to analyze transitive inferences syllogistically. The problems in this latter attempt will cry out for a logic of relations.

1. WHATELY AND THE REVIVAL OF FORMAL LOGIC

De Morgan refers to Richard (later Archbishop) Whately as the "restorer of logical study in England,"[1] an honor which seems well deserved. Whately's *Elements of Logic* was first published as a book in 1826, and it was often reprinted in the following years.[2] It was the first major treatise on *formal* logic to be written in England since Henry Aldrich's *Artis Logicae Compendium* of 1691. While other books on logic had been written in the intervening years, they were not primarily works on formal logic. For instance, Watts' *Logick* (1725) was most concerned with giving practical hints for effective reasoning, while Duncan's *Elements of Logick* (1748) emphasized Locke's psychological and epistemological theories.

The influence of Whately's book was due partly to its style, which Mary Prior has called "roughly Gilbert Ryle, vintage 1826."[3] Its historical importance does not lie in the rather traditional logical system it presents but in the viewpoint from which Whately presents it. Whately strongly emphasized the generality and importance of formal logic, and he conceived of logic as an abstract formal science. In these features he was closer to modern logic than to his immediate predecessors.

Formal logic had not been neglected in the preceding century for whimsical reasons; rather, it had been thought to be virtually useless. Against these claims, Whately defended both the utility and the theoretical importance of formal logic. He pointed out that criticisms of the utility of formal logic had been based on an exaggerated idea of what its utility ought to be. Formal logic does not seek a set of recipes for the acquisition and validation of knowledge. It has the more modest task of investigating the abstract patterns of argument which are common to all reasoning; and there should certainly be some utility in examining these.

Even if there is little practical value in formal logic, it may still furnish a type of theoretical understanding of humanity's peculiar activity–reasoning–and so legitimately be included in a liberal education. As Whately charmingly says:

> Even supposing that no practical improvement in argumentation resulted from the study of Logic, it would not by any means follow that it is unworthy of attention. The pursuit of knowledge on curious and interesting subjects, for its own sake, is usually reckoned no misemployment of time ...[4]

Whately's claims on behalf of formal logic must be born out by an examination of the actual results which formal logic can provide. It is not immediately obvious that reasoning, even of a deductive sort, has been limited to the types of reasoning normally encountered in formal logic. Whately meets this issue head-on by making two sweeping claims:

1. The validity of an argument must depend solely upon the logical forms of the propositions involved;
2. All arguments may be put into the form of the Aristotelian syllogism, and their validity assessed through applying the *dictum de omni et nullo*.

In the conjunction of these two theses we find the peculiar role of Whately's logic in the nineteenth century. His emphasis on the importance of logical form combines with other influences to create the increasingly formal logic of that period. At the same time, logicians such as De Morgan and Boole rejected his second thesis as they extended logic far beyond its Aristotelian borders. In the remainder of this section, we shall discuss the question of logical form; and in the next two sections, the adequacy of the Aristotelian syllogism.

Formal logic is supposed to deal with formal relations between propositions, yet the notions of *logical form* and *formal relations* are notoriously obscure. This is a problem on which Whately had very definite views and to which he devoted considerable attention. For him the proper task of logic was to,

> ascertain the validity or fallaciousness of any apparent argument, as far as the *form of expression* is concerned; that being alone the proper province of Logic.[5]

From this quotation it is clear that there is a close connection between the notion of logical form and the nature of the linguistic form in which arguments are expressed. This impression of Whately's method is confirmed if we follow the sequence of definitions which he uses to delimit the scope of logic: *Logic* is the Art and Science of Reasoning (p.25); *Reasoning* is an act of the mind which consists of proceeding from one judgment, to another founded upon that one (or the result of it) (p.69); an *Argument* is Reasoning expressed in words (p.84); and a *Syllogism* is an argument stated at full length and in its regular form (pp.84–85). Thus formal logic, which begins as the study of a type of mental act, becomes, in effect, the analysis of a certain kind of explicit linguistic structure, the syllogism. It should be emphasized, though, that these definitions do not imply that all syllogisms are Aristotelian in form; a syllogism is any fully explicit argument stated in its "regular" form.

From this preliminary set of definitions, Whately seems to think that a formalistic conception of logic follows immediately.

> Since, then, an argument is an expression in which *from something laid down and granted as true* (*i.e. the premises*) *something else* (*i.e.the conclusion,*) *beyond this must be admitted to be true, as following necessarily* (*or resulting*) *from the other*; and since Logic is wholly concerned in the use of language, it follows that a Syllogism (which is an argument stated in a regular logical form) must be "an argument so expressed, that the conclusiveness of it is manifest from *the mere force of expression*," i.e. without considering the *meaning of the terms*: e.g. in this syllogism, "Y is X, Z is Y, therefore Z is X," the conclusion is inevitable, whatever terms X, Y, and Z, respectively, are understood to stand for. And to this form all legitimate arguments may ultimately be brought.[6]

The crucial question, of course, is what he means by saying that "the conclusiveness of it is manifest from *the mere force of expression*, i.e. without considering the *meaning of the terms*." In the *Elements*, Whately leads the student to understand the importance of this "force of expression" in three different ways.

Whately begins his "Analytical Outline" by attempting to motivate the student through concrete examples. He points out that some arguments are either clearly valid or clearly invalid, but that others are more problematic, especially if "they are enveloped in a cloud of words" or if their conclusions are antecedently known to be true. In such cases, we should try to exhibit another argument which will "exactly correspond" with the given argument, but in which the premises are true and the conclusion is false. For instance, to use Whately's examples, the following argument might appear to be valid:

(1.1) Every rational agent is accountable.
Brutes are not rational agents.
———————————————
Brutes are not accountable.

However, to this argument there corresponds the argument,

(1.2) Every horse is an animal.
Sheep are not horses.
———————————————
Sheep are not animals.

The premises of this argument are true, but its conclusion is false.[7]

The crux of the issue, of course, is what is meant by saying that these arguments "exactly correspond." In some sense, yet to be determined, these two arguments have the same logical form; the logical skeleton is the same in both cases, and only the terms are different. A further specification of this logical form would require Whately's later analysis of the proposition into terms, copula and quantifier.

Whately remarks, however, that there is a second method which is more satisfactory than this method of counterexamples: we might construct a test which would provide necessary and sufficient conditions for the validity of an argument. Compared with the method of counterexamples, it appears that:

> to lay down such rules, and employ *them* as a test, is evidently a safer and more compendious, as well as a more philosophical mode of proceeding. To attain these it would plainly be necessary to analyse some clear and valid arguments, and to observe in what their conclusiveness consists.[8]

Whately then analyzes a typical Aristotelian syllogism in the first figure, and concludes that the *dictum* is the principle which makes it valid. He then makes the much stronger claim that "all valid arguments whatever may be easily reduced to such a form as that of the foregoing syllogisms; and that consequently the principle on which they are constructed is the UNIVERSAL PRINCIPLE of Reasoning."[9] Whately states this principle, the *dictum de omni et nullo*, in the following traditional terms:

> that whatever is predicated (i.e., affirmed or denied) universally, of any class of things, may be predicated, in like manner, (*viz*. affirmed or denied) of any thing comprehended in that class.[10]

This method involves a process of abstraction, for it requires that the argument be symbolized in a schematic way.

> In order to trace more distinctly the different steps of the abstracting process, by which any particular argument may be brought into the most general form, we may first take a syllogism stated accurately and at full length . . . and then somewhat generalize the expression, by substituting (as in Algebra) arbitrary unmeaning symbols for the significant terms that were originally used; the syllogism will then stand thus: "every B is A; C is B; therefore C is A." The reasoning is no less evidently valid when thus stated, whatever terms A, B, and C, respectively may be supposed to stand for; such terms may indeed be inserted as to make all or some of the assertions *false*; but it will still be no less impossible for anyone who *admits* the truth of the *premises*, in an argument thus constructed, to deny the conclusion; and this it is that constitutes the conclusiveness of an argument.[11]

Whately then points out that asserting the validity of all arguments of this form is just a verbally different way of stating the *dictum* itself. The main advantage in using these "arbitrary unmeaning symbols" is that we are not misled by the truth or falsity of the premises we are given, and we concentrate our attention instead on the connection between the premises and the conclusion.

> It belongs then exclusively to a syllogism, properly so called (i.e., a valid argument, so stated that its conclusiveness is evident from the mere *form* of the expression,) that if letters, or any other unmeaning symbols, be substituted for the several terms, the validity of the argument shall still be evident. Whenever this is not the case, the supposed argument is either unsound and sophistical, or else may be reduced (without any alteration of its meaning) into the syllogistic form; in which form, the test just mentioned may be applied to it.[12]

Invalid arguments cannot be reduced to such a form, but their invalidity may be shown most clearly by approaching the form as closely as possible, and then showing what is wrong.

The basic procedure for the determination of logical form, then, is to take the original argument and write out the equivalent syllogism by peering through the "cloud of words" to the crux of the argument, and then replace each of its terms uniformly by "unmeaning symbols."

So far, we have traced Whately's discussion of logical form through two stages: the method of counterexample and the method of the abstract schema. In the paragraph just quoted, a third element appears which is of considerable philosophical interest. To see this point, consider the following argument:

(1.3) John is a human being.

———————————

John is less than 200 years old.

To assess the validity of this argument, we must investigate the appropriate counterexamples (if any) or construct an appropriate schema. Suppose that we were to suggest that its schema is,

(1.4) x is a human being.

———————————

x is less than 200 years old.

Whately, of course, would not admit this as an appropriate schema, since not all terms have been replaced by unmeaning symbols. The proper schema is, rather,

(1.5) x is A.

———————————

x is B.

which is patently invalid. What lies behind Whately's insistence that (1.5) is the proper schema, rather than (1.4)? There are at least two ways of supporting Whately's conception of schematization.

To begin with, it is plausible to say that (1.3) is a misleading way of stating an argument that would be better stated as:

(1.6) All human beings are less than 200 years old.
John is a human being.

———————————

John is less than 200 years old.

There must be some distinction between the task of the logician and that of the scientist: the logician works solely within the domain of what follows from what, while the scientist is concerned with facts about the world. If (1.4) were taken as a perspicuous argument, to be assessed by the logician, we would be forced to conclude that there are many arguments regarding whose validity the logician must remain silent, since their validity involves matters of empirical fact. It is clarifying to separate empirical and logical issues by putting all empirical assumptions into the premises, and then considering the relation between premises and conclusion as purely a matter of logic. When we see that (1.3) is invalid as stated, we may then search for a missing true premise which would render it not only valid but sound.

At this point, it should be noted that the methods of counterexample and schematization do not eliminate the notion of necessity in the way in which similar accounts may attempt to do today. An argument is often said to be valid if there is no interpretation of its terms which renders its premises true and its conclusion false. This may appear to clarify the vague notion of necessity by replacing it with the clearer notion of the instances of an argument form. Whately, however, did not think of the matter in this way. He thought of the use of counterexamples and schemas as practical tools, not as means of philosophical clarification. His remarks have made it clear that the conclusiveness of an argument is evident in and of itself, when it is stated in proper form. The methods of counterexample and schematization do not constitute the grounds for the validity of the argument; they serve only as ways of clearing the mind of confusions that might arise from literary style or from antecedent beliefs about the truth or falsity of the premises or the conclusion.

We may observe, then, that the method of schematization shows that we are concerned only with conclusions that follow necessarily from their premises, not those in which empirical facts are at issue. Premises must be added, when necessary, so that the full logical form can be exhibited. The proper schema for (1.6), and thus for (1.3), will be,

(1.7) All A are B.
x is A.
___
x is B.

However, there is an additional step required before we can justify the claim that Whately's method of schematization is the required method. We must show that there is some connection between the necessity by which a conclusion follows from its premises, and the use of a schema in which terms are replaced by unmeaning symbols, while logical symbols retain their ordinary meanings. We must determine, that is, the reasons why Whately claims that in a properly formulated valid argument, "the conclusiveness of it is manifest from *the mere force of expression*, i.e. without considering the *meaning of the terms*."[13] For we have now seen that the form of the argument is to be found in "the mere force of expression" and not in the meaning of the terms.

Unfortunately, the basis for Whately's claim is not completely clear. To some extent it seems implicit in the traditional manner in which he builds up his analysis of the proposition, beginning first with terms, and then showing how they may be combined into propositions through the copulas and quantifiers. More deeply, however, it appears that he may be reasoning in the following manner: logic is concerned solely with the use of language; thus, it must be concerned solely with the linguistic devices that provide the structure of the syllogism, and not with the content of any of the terms that make up a syllogism. If non-logical terms were to remain in the schemas, the logician would be forced to know something about those terms, so that the validity would be not just a matter of language but a matter of fact. Only facts about the behavior of logical expressions can be within the purely linguistic domain of the logician.

This discussion of Whately's views on logical form raises most of the issues that are required to place De Morgan's views on logical form in context. At this point, however, we must ask whether all validity is actually to be found in the behavior of the logical constants. Consider, for instance, the following argument:

(1.8) John is taller than Harry.
Harry is taller than Bill.
―――――――――――――
John is taller than Bill.

The conclusion certainly follows necessarily from the premises, but this cannot be shown by a typical syllogistic schema. What this suggests, of course, is that the argument lacks a premise, and that it should read:

(1.9) If one thing is taller than a second and the second is taller than a third, then the first thing is taller than the third.
John is taller than Harry.
Harry is taller than Bill.
―――――――――――――
John is taller than Bill.

This allows us to maintain that the logician deals only with "the mere force of expression," and not with the meanings of terms. We must realize, however, that the needed additional premise may itself be a necessary truth and not an empirical truth, even though both types of premises embody knowledge which the logician does not possess *qua* logician. We may still wonder, of course, whether (1.9) is an instance of a valid Aristotelian schema, for it is not obviously of this sort. This raises the more general question of the adequacy of the Aristotelian syllogism, which is the subject of our next two sections.

2. EUCLID AND THE SYLLOGISM

Nineteenth century English logic exhibits several attempts to break out of the straightjacket of the Aristotelian syllogism. It is ironic that Whately's own position on this issue is extremely conservative. He believed that all arguments may be adequately symbolized in the Aristotelian syllogism, and that their validity may be fully assessed through the *dictum*. In order to make this thesis plausible, he must take heroic measures with apparent counterexamples. For instance, *modus ponens*,

(1.10) If p, then q
p
―――――
q

is transformed into

(1.11) The case of p is a case of q.
This is a case of p.
―――――
This is a case of q.

where p and q can be absolutely specific singular propositions.[14] Inductive arguments are transformed into:

(1.12) What belongs to the individuals we have examined belongs to all individuals of the class to which they belong.
P belongs to all the individuals we have examined.
―――――
P belongs to all individuals of the class to which they belong.[15]

Rather than pursue the problems raised by these transformations, we shall pursue other problems more relevant to the development of the theory of relations: the possibility of syllogistically analyzing those arguments which depend upon the transitivity of certain relations, such as *equality*, *greater than*, and *taller than*. This is not a mere technical exercise, for it goes to the heart of the issue of the utility of syllogistic logic. Mathematical reasoning, both in geometry and in arithmetic, makes continual use of these arguments. If syllogistic logic is not able to handle them, it must be inadequate for the most impressive examples of demonstrative arguments.

For over two thousand years, Euclidean geometry was justifiably viewed as the supreme example of demonstrative reasoning. As a demonstrative science, geometry uses rigorous logic to generate innumerable truths from a few evident first principles. During that same two thousand years, some type of syllogistic logic was commonly thought to be adequate for handling all correct deductions, even if it meant supplementing Aristotle's logic with some propositional logic such as the hypothetical syllogism.

Amazingly, though, these two millennia saw virtually no interest in the question of whether syllogistic logic sufficed for Euclid's inferences. As Ian Mueller has shown, the origins of Greek mathematics and of Greek logic owe almost nothing to each other.[16] Furthermore, until the late nineteenth century, the history of geometry shows no serious attempts to construe Euclidean geometry syllogistically, or to examine in detail *even one* Euclidean proof to see whether it could be treated syllogistically. When logicians or mathematicians did approach these issues, they did so in a cursory way and with little sense of the enormity of the issues at stake for logic and the philosophy of geometry. Even in the late nineteenth century, when modern symbolic logic was being developed, it was number theory, set theory and analysis that provided the stimulus, and not geometry. And all of this is so, despite the fact that geometry was the paradigm of demonstrative science for two thousand years.[17]

This ironic *independence of logic and geometry* presents a very disturbing historical puzzle. How could generations of logicians claim that the syllogism is the universal canon of valid deduction, when they could not be sure that it suffices for even the first theorem of Euclid? Or, if they knew that there are some difficulties in doing this, why did they not see the significance of this fact and pursue the matter further?

To see some of the more serious problems, let us consider Proposition I of Euclid's First Book.

Proposition I.

On a given finite straight line to construct an equilateral triangle.

Let AB be the given finite straight line. Thus it is required to construct an equilateral triangle on the straight line AB.

With centre A and distance AB let the circle BCD be described; [Post. 3]
again, with centre B and distance BA let the circle ACE be described; [Post. 3]
and from the point C, in which the circles cut one another, to the points A, B let the straight lines CA, CB be joined. [Post. 1]

Now, since the point A is the centre of the circle CDB,

AC is equal to AB. [Def. 15]

Again, since the point B is the centre of the circle CAE,

BC is equal to BA. [Def. 15]

But CA was also proved equal to AB;
therefore each of the straight lines CA, CB is equal to AB.

And things which are equal to the same thing are also equal to one another; [C.N.1]

therefore CA is also equal to CB.

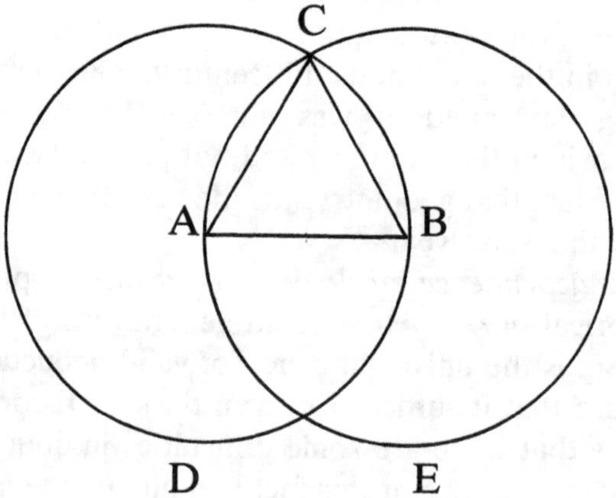

Therefore the three straight lines CA, AB, BC are equal to one another.

Therefore the triangle ABC is equilateral; and it has been constructed on the given finite straight line AB.

(Being) what it was required to do.[18]

The basic idea behind this proof is simple; but its logical formulation is complex. For one thing, the Proposition is not a "theorem" in a straightforward sense, but, rather what later came to be called a "problemata"–that is, a set of instructions for performing a task. The first real "theorem" in Euclid is Proposition 4 of Book I. Furthermore, the proposition requires that several constructions be drawn, and it is not clear how these should be represented logically. Constructions are also required for the proof of many theorems. In addition, Euclid did not include a complete set of common notions and postulates, so that additional principles must be added; for Proposition I, for instance, we must assume that the two circles do meet at some point.

When all of these problems have been solved, we are still faced with the formidable task of logically analyzing the crucial inferences in the proof. Few partisans of syllogistic logic seem to have taken this problem seriously. Sir William Hamilton, though, refers to one attempt to carry out this task.

> Leibniz (*Opera*, t.ii.p.17) commemorates the notable exploit of two zealous, but thick-headed logicians–Herlinus and Dasypodius by name–who actually reduced the first six books of Euclid into formal syllogisms.[19]

Since De Morgan also refers to the work of Herlinus and Dasypodius as the one attempt to construct syllogistic proofs of Euclidean theorems, it will be useful to consider their attempt in some detail.[20]

Herlinus and Dasypodius distinguish between the various parts of a proof in the traditional way; we shall focus on the "apodeixis," which constitutes the core of the proof. They remark:

> Apodeixis itself at times consists of many arguments or means of syllogisms. Therefore, in order to remove obscurity and difficulty, the propositions of syllogisms are separated, and what were connected by Theon with acute brevity are here more widely spread and more openly rendered.[21]

All of this was done for pedagogical reasons, "for the benefit of good young people, ... at least with benevolence toward students of geometry." Dasypodius realizes that this way of rendering these proofs may seem artificial.

> Perhaps there will be those who will say that these are too scholastic and puerile. I confess and say openly and candidly that I write these things for boys and for our students who out of classes after the learning of languages and also rhetoric and dialectic are led down to geometry and descend into the arena of geometers that they may exercise themselves there and strengthen their judgment which is as yet tender.[22]

The primary purpose of this "syllogistic" rendering of Euclid, then, is not theoretical but pedagogical, making Euclid more accessible to students who have already learned their syllogistic logic in studies of dialectic.

How, then, do Herlinus and Dasypodius construct this proof? Here is a translation of their proof, along with explanations which are compressed versions of their "Explicatio."[23]

To Prove: Triangle ABC is equilateral.

Proof	*Explanation*
1. In every circle, straight lines from the center to the circmferance are equal.	Def. of "circle"
2. Figure BCD is a circle whose center is A.	By construction
3. Therefore, line AC is equal to line AB.	From 1, 2
4. In every circle, etc., as above.	Def. of "circle"
5. Figure ACE is a circle whose center is B.	By construction
6. Therefore, line CB is equal to line BA.	From 4, 5
7. Things equal to the same thing are equal to each other.	Common notion
8. Each of the straight lines AC, CB is equal to the straight line AB.	3 and 6
9. Therefore, line AC is equal to line CB.	7, 8
10. Whatever triangle is contained by three equal straight lines is an equilateral triangle.	Def. of "equilateral triangle"
11. Triangle ACB is contained by three equal straight lines.	"The conclusion of the third syllogism"
12. Therefore, triangle ACB is equilateral.	10, 11

This is a sequence of four arguments. It certainly has a syllogistic feel about it, for each argument contains two premises and involves moving from the general to the particular. Quite clearly, though, none of the first three arguments are Aristotelian syllogisms, for each has more than three terms. The first two arguments (Steps 1–3 and 4–6) are actually rather difficult, for the complex definition of a circle means that the inferences which are needed to extract its content are quite complicated, even though Euclid hardly mentions them. The third argument (Steps 7–9) involves the transitivity of equality, an issue which we shall pursue in the next section. Only the fourth argument (Steps 10–12) is traditionally syllogistic.

Given the evident complexities of this attempt to "syllogize" Euclid, it is remarkable that anyone should think that it was successful. Even Leibniz, to whom Hamilton appears to owe the reference to Herlinus and Dasypodius, seems to have accepted their claim to have represented Euclidean demonstrations syllogistically, even though he recognizes that the traditional syllogism may not exhaust all the logical forms.

> A demonstration is sound when it observes the form prescribed in logic, although it need not always follow the form of syllogisms arranged in the Scholastic manner (such as Christian Herlinus and Conrad Dasypodius applied to the first six books of Euclid); it is merely necessary that the argument be conclusive by virtue of its form.[24]

Unfortunately, Leibniz does not indicate what other forms he has in mind. Ian Mueller has shown, though, that this proof of Euclid's first proposition requires the full resources of first-order quantification theory, with one-, two- and three- place predicates. Instantiation, conditionalization and generalization are used.[25]

3. REID, HAMILTON AND MANSEL ON RELATIONAL INFERENCES

Let us focus our attention on the third argument in Herlinus and Dasypodius' reconstruction of Euclid. In De Morgan's time, the sufficiency of the syllogism for mathematical proof had been questioned by Thomas Reid on the ground that it could not handle such inferences as,

(1.13) A is equal to B.
B is equal to C.
―――――――――
A is equal to C.

This is the core of the argument which constitutes the third step, though other principles are involved as well.

Thomas Reid's criticisms of syllogistic logic appear in his *A Brief Account of Aristotle's Logic*, which became generally available when it formed part of Reid's *Works* which were edited and annotated by Sir William Hamilton.[26] Reid did not have a very high opinion of syllogistic logic–not because he thought it to be wrong, but because he believed that it applied to a very limited domain of reasoning. He argued that it did not apply to most mathematical reasoning, the one area in which our reasoning power is best exhibited. His reasons for making this claim are interesting and clear.

> But it is to be observed, that the propositions of mathematics are not categorical propositions, consisting of one subject and one predicate. They express some relation which one quantity bears to another, and on that account must have three terms. The quantities compared make two, and the relation between them is a third. Now, to such propositions we can neither apply the rules concerning the conversion of propositions, nor can they enter into a syllogism of any of the figures or modes. We observed before that this conversion *A is greater than B*, therefore *B is less than A*, does not fall within the rules of conversion given by Aristotle or the Logicians; and we now add, that this simple reasoning, *A is equal to B*, and *B to C*, therefore *A is equal to C*, cannot be brought into any syllogism in figure and mode.[27]

It will be useful to analyse this passage in some detail, and to discuss three crucial aspects of it. In each case we shall contrast Reid's position with that of his editor, Sir William Hamilton, who liberally sprinkled Reid's *Works* with corrections in the form of footnotes.

The first point to note is that Reid's most fundamental thesis concerns the analysis of relational propositions, for he claims that they really have three terms, not two. According to Reid, relational propositions differ essentially from categorical propositions, in that the latter are of the subject-predicate form while the former involve two terms and a relation. This is the ultimate source of the fact that the Aristotelian syllogism cannot be successfully applied to inferences which involve them. Hamilton, in a footnote, simply denies Reid's thesis by saying, "These propositions are categorical," without any further explanation. Presumably, though, he would claim that "A is greater than B" is of the subject-predicate form, with "greater than B" as the predicate.

Keeping in mind our previous comments on the logic of geometry, the second point of interest is Reid's assertion that inferences involving the transitivity of equality cannot be treated by the traditional syllogism. The slight treatment given by Reid to this point suggests that he was not aware of its dramatic implications for the whole of logical theory and, indeed, for the philosophy of geometry; but Hamilton may well have seen what was at stake, for he defended the adequacy of the syllogism in the face of this argument. In a footnote to the passage quoted above, Hamilton added:

Not as it stands; for, as expressed, this reasoning is elliptical. Explicitly stated it is as follows:–

What are equal to the same are equal to each other;
A and C are equal to the same (B);
Therefore, A and C are equal to each other.

He then refers approvingly to the book by Herlinus and Dasypodius, "in which every demonstration is developed in regular syllogisms." Two footnotes later, he adds, in ringing terms, " 'The compass of the Syllogistic System' [Reid's phrase] is the compass of the reasoning faculty of man."[28]

There is much to be said for Hamilton's position on this inference. For one thing, it surely reflects Euclid's own practice, for Euclid's proof of Proposition I explicitly invokes Common Notion I, which reads, "Things which are equal to the same thing are equal to each other." But let us apply Hamilton's strategy to Euclid's proof and see if it will work. The argument would then be:

(1.14) Things equal to the same thing are equal to one another.
CA and CB are things equal to the same thing.
___
CA and CB are equal to each other.

Already we must be a bit suspicious. It is not at all clear that the minor premise and the conclusion are categorical propositions at all, for each asserts that a certain pair–the pair *CA and CB*–has a certain property. It seems more natural just to say that CA is equal to CB; and it is not clear whether, say, traditional logic would countenance relations as properties of pairs. But let us put aside these qualms, and see if this strategy will work formally. In fact, as Mueller shows, it will not.[29] The minor premise of (1.14) must be inferred as follows:

(1.15) CA and AB are equal to one another.
CB and AB are equal to one another.
___
CA and CB are equal to the same thing.

This is not a categorical syllogism, for it has five terms. Hamilton's formulation would only mask the problematic character of this inference by writing the original minor premise as,

CA and CB are things equal to the same thing (AB),

where there is no indication that their equality to the same thing must be *inferred* from their equality to AB. He might reply, of course, "That is obvious," but the point of logic is to capture the obvious.

There is a different strategy which might be used to establish the needed minor premise. One might argue,

(1.16) What are equal to AB are equal to the same thing.
CA and CB are equal to AB.
____
CA and CB are equal to the same thing.

This, however, is dubious on several fronts. For one thing, we would need to argue,

(1.17) CA and AB are equal to one another.
CB and AB are equal to one another.
____
CA and CB are equal to AB.

The conclusion of this argument is just equivalent to the conjunction of its premises, but how is this to be represented categorically? Furthermore, the complex subject in the conclusion functions like "John and Mary" in "John and Mary are tall," while in the premises the complex subjects involve a genuine relation.

More importantly, the major premise in (1.16) shows that something is wrong. For one thing, it is specific to AB, so that a full axiomatization of geometry which used such principles would require a separate principle for each such inference, or at least a schema. Most basically, this major premise should not be required at all, since it is not a material principle of geometry (or even of the theory of equality), but a logical principle. It is merely the propositional form of an instance of Existential Generalization, so it should not be required as an extra premise. We may take it as a sign of defective analysis when logical principles must be added as premises to secure the reduction of an argument to some preferred form. Otherwise, we could insure the reduction of any argument to *modus ponens* by taking as our major premise a conditional, whose antecedent is the conjuction of its premises and whose consequent is its conclusion.

Equality is not the only relation whose transitivity presents a problem. Mansel, for instance, considers the "a fortiori" inference,

(1.18) A is greater than B.
B is greater than C.

A fortiori, A is greater than C.

He points out that we might mean by the conclusion that A is even greater than C than B is; but that if we do not want to include this in the force of the conclusion, we can formulate it as follows:

(1.19) Whatever is greater than a greater than C is greater than C.
A is greater than a greater than C.

A is greater than C.[30]

While this formulation is rather odd, the argument can be restated as.

(1.20) Whatever is greater than something which is greater than C is greater than C.
A is greater than something which is greater than C.

A is greater than C.

This strategy, which is so similar to Hamilton's, shows the same problem as Hamilton's treatment of equality: somehow, the "B" has been lost. To obtain the minor premise of (1.20), we would need,

(1.21) A is greater than B.
B is greater than C.

A is greater than something which is greater than C.

This is hardly a traditional syllogism, since it has five terms.

This example also shows that the treatment of equality, unsuccessful as it was, was artifically easy, because of the special properties of equality. There is no easy way of concocting a new general major premise for (1.21) as we did for (1.16), since *greater than* is not a symmetrical relation. The only remaining option would be to throw up our hands and just formulate the argument as,

(1.22) If A is greater than B and B is greater than C, then A is greater than something which is greater than C.
A is greater than B and B is greater than C.

A is greater than something which is greater than C.

Whately's method might then be used to convert this hypothetical syllogism into a categorical syllogism. At this point, though, we must note that the syllogistic thesis has once again been trivialized through adopting a major premise which is just a logical principle; and, as in the case of equality, this shows that the analysis is defective.

One final way in which to look at transitivity arguments is to note that the real problem arises from the natural insistence that a general principle be used as the major premise. If this were not insisted upon, Mansel and Hamilton could just take (1.22) one step further and formulate the argument as follows:

(1.23) If A is greater than B and B is greater than C, then A is greater than C.
A is greater than B and B is greater than C.
―――――――――――――――――――――
A is greater than C.

This, also, could be converted to a categorical syllogism by Whately's technique; and the same strategy could be applied to equality.

This might appear to trivialize the thesis of syllogistic reduction, but the situation is less clear than it was with the major premises of (1.16) and (1.21). The major premise of (1.23) differs from the other two cases in not being a logical or formal principle, but rather a material principle about the relation of *greater than*. It is thus legitimate that it be added as a premise to bring the argument into proper form for logical evaluation. The same would be true for the transitivity of equality. Mansel seems to recognize this possibility, when he says:

> Moreover, even where true premises can be added, and the Consequence legitimately deduced, we cannot, except from knowledge of the matter, determine into what form the reasoning will naturally fall. In some cases, . . . the proof may be given in Categorical Syllogisms. In others, it is far more naturally exhibited in the hypothetical form. The assumption of a hypothetical premise will always present the reasoning in a logically valid, though not always in the most natural, form; and it is frequently the only materially allowable assumption in cases where the given antecedent and consequent have both terms distinct. E.g. A is B, therefore C is D. We may supply, if A is B, C is D; but to determine the truth of the assumed proposition, whether it be hypothetical or categorical, does not fall within the province of the Logician.[31]

The main objection to (1.23), and its analogue for equality, is that it does not really carry out the intentions of those who insist on the primacy of the syllogism; and, in fact, it does not represent Euclid's reasoning as it actually occurs. Hamilton is correct in saying that if a premise is missing from (1.13), it is Euclid's first Common Notion, which is a general statement about equality. If a premise is added to (1.18), it should be a general statement of the transitivity of *greater than*. Such a principle is more difficult to formulate for *greater than* than for equality, and it would seem to involve something like a covert appeal to variables and quantifiers. It might be stated as,

(1.24) If one thing is greater than a second thing, and the second thing is greater than a third thing, then the first thing is greater than the third thing.
A is greater than B and B is greater than C.

A is greater than C.

This is not a traditional syllogism, yet it seems closely related to it. The crux of the traditional syllogism is the application of general principles to particular cases; that is *subsuming* an individual, or a class of individuals, under a general principle. This was traditionally limited to one subsumption at a time, in the subject term. The inference (1.24), on the other hand, may be construed as a case of *multiple subsumption*, applying a universal proposition to particular cases of each of the terms which occur in it. One might attempt to support Hamilton's general standpoint by extending the traditional syllogism in a conservative manner to include inferences such as (1.24). One might wonder, though, whether this move is as conservative as it appears and ask whether this is the logic of relations in disguise.

We have now discussed two of the three issues raised by Reid: the relational analysis of propositions and the treatment of inferences which depend upon the transitivity of a relation. The third issue which he raises may be treated more quickly than the second.

Reid points out that the syllogism is also not applicable to such inferences as.

(1.25) A is greater than B.

B is less than A.

Such inferences, he notes, do not "fall within the rules of conversion given by Aristotle or the Logicians." In an earlier passage, he had said that "we are left to the direction of common sense" in cases such as (1.25) and,

(1.26) Alexander was the son of Philip.

Philip was the father of Alexander.

We are dealing here, of course, with relations and their converses; and, since "father" is not the converse of "son," the point of (1.26) could be better stated as,

(1.27) Alexander was a child of Philip.

Philip was a parent of Alexander.

Hamilton's footnotes provide the obvious response to this claim: these propositions are categorical propositions, and the reason why they are beyond the rules of logic is that, "they are beyond the sphere of logic, being *material* not *formal* conversions."[32] While Hamilton does not state his criterion of materiality, he presumably means that the validity of such an inference does not depend merely upon its form, but upon the particular relations involved. Mansel agrees when he says that, "in both these it is our material knowledge of the relations 'father and son,' 'greater and less,' that enables us to make the inference."[33]

In line with Hamilton's and Mansel's general approach to material assumptions, this knowledge would best be rendered by an additional general principle, and the resulting inference treated as a traditional syllogism. In doing so, though, we encounter the same problems which arose in the latter part of our treatment of transitivity. While (1.25), for instance, could be completed by adding the major premise, "If A is greater than B then B is less than A," this would not capture our sense that a general principle should be invoked in this case. Providing such a general principle would seem to require the same strategy as was used in (1.24), obtaining,

(1.28) If one thing is greater than a second thing, then the second thing is less than the first thing.
A is greater than B.

B is less than A.

Once again some type of multiple subsumption would seem to be required to validate this argument. And, while not strictly syllogistic, it would be syllogistic in spirit.

This case, however, may raise some doubts about the distinction between the formal and the material on which our discussion of missing premises has been based. The basic thesis is that the addition of new material principles is allowable, since material knowledge should be represented by the premises of an argument, but that the logical analysis of an argument is defective if it requires the inclusion of a formal principle as a premise. We must ask, then, whether the major premise of (1.28) is a material or a formal principle. It might be interpreted to mean only that if x bears R to y, then y bears R-converse to x; under this interpretation, it only reformulates the principle behind the original inference and is not a material principle at all. Perhaps, though, this added premise embodies the claim that "less than" expresses the converse of "greater than"; but whether this should be considered material knowledge, or whether it should be considered only a matter of the meanings of words, is unclear. In a formal system, of course, one might just define "less than" as the converse of "greater than"; but the relevance of this to ordinary English is also unclear.

Obscurity about the distinction between the formal and the material might make us uncertain about the examples of transitivity which we have just considered. Are the transitivity of *greater than* and *equals* part of the very meanings of those terms, or are they more substantive principles? Perhaps more drastically, are "greater than" and "equals" both descriptive terms, or do they somehow represent the form of the proposition? If the answer seems obviously to be the former, what about the identity sign? De Morgan will force us to return to these issues later.

Before beginning our consideration of De Morgan's views on these subjects, let us note that we have only considered a small sample of the many mathematical inferences which an adequate logic must handle. Mansel, in fact, considers two other types of cases in attempting to establish his claims for the adequacy of the syllogism. In one of these, he discusses the following interesting argument:

(1.29) A is equal to B.

Twice A is equal to twice B.

which he reconstructs as,

(1.30) The doubles of equals are equals.
Twice A and twice B are doubles of equal things.

Twice A and twice B are equal.[34]

The introduction of functions into the discussion results in complexities which cloud the analysis of the argument. However, we may notice that the same issues arise here which we have already considered. Mansel is forced to take "twice A and twice B" as a complex logical subject; and he must provide some way of moving from "A is equal to B" to "twice A and twice B are doubles of equal things." While these latter two propositions are undoubtedly equivalent, the logician's question is, why? Once again, multiple subsumption seems the best way out:

(1.31) If one thing is equal to a second thing, then twice the first thing is equal to twice the second thing.
A is equal to B.

Twice A is equal to twice B.

As before, it is also not clear whether the major premise of (1.31) is a material or a formal principle. If equality between numbers is taken to be identity, then it seems to be but an instance of the indiscernibility of identicals.

In a separate section on the Demonstrative Syllogism, Mansel implicitly contradicts Reid by saying that geometrical argument is the example *par excellence* of the demonstrative syllogism.[35] He notes that since Aristotle's theory of demonstration is designed primarily for geometry, it would be odd indeed if it were found not to apply to that science. To establish his claim, he provides the following example of a geometrical inference.

(1.32) Magnitudes equal to the adjacent exterior and interior angles of a triangle are equal to each other.
The three interior angles and two right angles are equal to the adjacent exterior and interior angles.

The three interior angles and two right angles are equal to each other.

This appears to be a straightforward syllogistic inference, but its major premise is supported by an inference which raises issues which are familiar by now.

(1.33) Magnitudes which are equal to the same are equal to each other.
Magnitudes equal to the adjacent exterior and interior angles are equal to the same.

Magnitudes equal to the adjacent exterior and interior angles are equal to each other.

In the last section we noted that few efforts were made to test seriously the thesis that the traditional syllogism embodies the logic of Euclid's *Elements*. We have now seen that Thomas Reid suggested that inferences involving the transitivity of relations, as well as the conversion of relations, showed the uselessness of syllogistic logic in dealing with many mathematical inferences. Hamilton's and Mansel's attempts to solve this problem are beset by several difficulties of which they were apparently not aware; and it appears that the only way of resolving these problems in a syllogistic spirit is to allow for some type of multiple subsumption.

With this background in mind, we are now in a position to consider the early stages of De Morgan's thinking about relations.

CHAPTER II

FIRST THOUGHTS ON THE COPULA

Throughout his life, De Morgan's work in logic showed an intriguing combination of conservatism and novelty. Thus, he wrote very extensively on the traditional categorical syllogism, yet he treated this subject in a variety of original ways. De Morgan's approach to relations and relational inferences showed this same mixture of novelty and tradition. This is especially true of De Morgan's early writings on logic, from 1831 to 1843, which are the subject of this chapter. Three main issues will attract our attention.

In his two earliest writings on logic, De Morgan makes strong claims for the adequacy of the traditional syllogism. Yet, in the first of these writings, he is able to accommodate mathematical reasoning only by allowing *is equal to* as a copula, in addition to the usual *is*. This can be viewed as the origin of his later attempts to construct a generalized theory of the copula.

In these same writings, he recognizes the obvious importance of relational arguments, especially *a fortiori* arguments–that is, arguments such as, "A is less than B; B is less than C; therefore, A is less than C." Yet, he struggles mightily to reduce these inferences to the categorical syllogism.

Finally, in all these early writings, he describes inference in highly general relational terms, such as, "the establishment of one relation as a necessary consequence of others." The generality of this account is immediately subverted, though, by the claim that the essential element in any relational proposition is the identity relation. In De Morgan's later writings, the generality of his relational account of inference will gradually displace his attachment to the identity relation.

1. THE TWO COPULAS

In 1828, Augustus De Morgan was appointed Professor of Mathematics at the University of London (now University College) at the age of twenty-two, winning the post over some thirty other candidates. His Introductory Lecture, delivered at the opening of classes on November

5, 1828, was a strong defense of the teaching of mathematics, along with comments on how he proposed to do it. De Morgan stressed the importance of mathematics for the general student as a way of inculcating sound habits of reasoning. Mathematics, he claimed, has special virtues for this purpose.

It is obvious that whatever be the subject which is selected as the instrument by which the art of reasoning should be taught to those unaccustomed to demonstration, it should if possible be free from extraneous difficulties, that is from difficulties which are independent of the reasoning itself.[1]

The clarity of geometry makes it ideally suited for this purpose, since pure geometrical ideas are the "clearest and most definite which our minds contain."

De Morgan then provides a very attractive account of the traditional conception of geometry, which he strongly supports.

I come now to the first principles which must be admitted before any attempt to reason upon the nature of the objects to which names have been given can be attended with success. An edifice may indeed be erected by the skill of the builder but the soil under the foundation, must be rendered secure by the hand of nature, and adapted to the purpose by that of man, or no stability can be expected. If reason be the art of deducing complex results from those which are simple by satisfactory demonstration, and if a science be, as it has been defined to be, a systematic body of knowledge whose parts are connected by sound reasoning, there can be no science which has not some undeniable truths as its groundwork, some propositions which are too simple to be either established or destroyed by demonstration.[2]

Geometry, then, presents us with unambiguous terms, self-evident first principles, and conclusive demonstrations. It would be hard to find a clearer endorsement of the Euclidean ideal of knowledge which has played such a role in western epistemology.

All of these points are reiterated in the introductory chapter of *On the Study and Difficulty of Mathematics*, which he published three years later. This makes it important for him to address the difficulties in mathematics–not just the stumblings of the elementary student or the pedagogical problems of the teacher, but the genuine conceptual problems in the mathematics of his day, such as the use of negative and "impossible" quantities. It also means that he must give some account of the logical structure of geometrical reasoning, to which he devotes an entire chapter. On more than one occasion he emphasizes that there is nothing esoteric about the laws which govern geometrical reasoning, for "the commonest actions of our lives are directed by processes exactly identical with those which enable us to pass from one proposition of geometry to another" (SDM, 230).

De Morgan's treatment of geometrical reasoning is rather conservative, for he appears to endorse the claim that the traditional syllogism is adequate for it. He presents a fairly traditional account of the forms of categorical propositions, and he outlines the traditional moods and figures of the syllogism and presents the syllogistic rules. After presenting the traditional system, he concludes that,

> The elements of geometry present a collection of such reasonings as we have just described, though in a more condensed form. It is true that, for the convenience of the learner, it is broken up into distinct propositions, as a journey is divided into stages; but nevertheless, from the very commencement, there is nothing which is not of the nature just described.(SDM, 220)

To illustrate this fact, he then produces a proof of the Pythagorean Theorem which has been reduced "nearly to a syllogistic form," with only some "petty" inferences omitted.

If De Morgan had continued to hold this conservative position he would never have become an important figure in the history of logic. With this in mind, it is important to note that underneath this apparently conservative analysis, there are some deviations from the logical tradition which contain the germ of De Morgan's later ideas. This can be seen both in his analysis of the proposition and in his treatment of inference. On the propositional level, he claims that,

> Every assertion can be divided into three distinct parts. Thus the phrase, "all right angles are equal," consists of: (1) the *subject* spoken of, viz., right angles, which is here spoken of universally, since every right angle is a part of the subject; (2) the *copula*, or manner in which the two are joined together, which is generally the verb *is*, or *is equal to*, and can always be reduced to one or the other: in this case the copula is affirmative; (3) the *predicate*, or thing asserted of the subject, viz., equal angles.(SDM, 203)[3]

He claims that all assertions are of one of the four traditional forms, even though significant changes may be needed to reduce an assertion to one of these forms–e.g., "parallel lines never meet" must be reduced to "parallel lines are lines which never meet."

This analysis of the proposition allows him to assert that there are two quite different forms of each syllogistic mood and figure. In 1 AAA, for instance, there is both,

(2.1) All A are B.
x is A.
―――――――
x is B.

and,

(2.2) A is equal to B.
B is equal to C.
———————
A is equal to C.

It is interesting to note that De Morgan nowhere comments on the logical novelty of this *doctrine of the two copulas*, as we may call it, nor on the novelty of its use in inference.

There is no evidence to suggest that De Morgan was aware in 1831 of the controversies surrounding the transitivity of equality. His early work contains no references to Reid's criticisms. Mansel's edition of Aldrich had not appeared by this time, and Hamilton's edition of Reid was not to appear until 1846. Nevertheless, it will be useful to contrast De Morgan's views with those of Reid and Hamilton.

According to Hamilton's conservative view, there is only one affirmative copula, *is*. This is the copula in "A is equal to B" just as much as in "A is B." The only difference is that in the former case, the predicate is a complex predicate, "equal to B." In order to validate (2.2), a new major premise must be provided, asserting the transitivity of equality. We have seen the technical problems which utilizing such a premise presents. Reid, on the other hand, refuses to construe "A is equal to B" as a categorical proposition; he insists that it is of a different form, a relational form. "Is" is still its copula, but the subject, the predicate and the relation make three terms. This contrasts dramatically with categorical propositions of the form "A is B," which have only two terms in addition to the copula. Similarly, Reid refuses to reduce arguments involving the transitivity of equality to syllogistic form. While he says that the syllogism cannot be used to validate them, he does not specify what form the rule governing them would take. Nevertheless, it seems reasonable to suppose that he would use a primitive rule of inference to validate such inferences as (2.2), just as a syllogistic rule validates (2.1). The rules would be on exactly the same level, with neither being reduced to the other.

De Morgan agrees with neither Hamilton nor Reid on the analysis of these two propositional forms. Against Reid, he asserts that they have the same logical form: subject-copula-predicate. It is just that there are two copulas, *is* and *is equal to*. This contrasts with Hamilton's claim that there is only one copula, *is*, and that "equal to" forms part of a complex predicate. We may contrast the three analyses of both propositions in the following way:

	A is B			A is equal to B			
	subject	copula	predicate	subject	copula	relation	predicate
Hamilton	A	is	B	A	is		equal to B
Reid	A	is	B	A	is	equal to	B
De Morgan	A	is	B	A	is equal to		B

How, Reid would ask, can De Morgan think of both assertions as being of the subject-copula-predicate form, when there is mere predication in one case and a relation in the other? A likely hypothesis is that De Morgan can say this because even at this stage he is giving a relational analysis of *both* copulas. In A is B, "is" expresses one type of relation; and in A is equal to B, "is equal to" expresses another relation, no more, but no less, fundamental than the first.

On the inferential level, De Morgan would agree with Reid, against Hamilton, that the logical rules for "is" and "is equal to" are equally fundamental, and he totally rejects Hamilton's claim that (2.2) requires an additional material premise asserting the transitivity of equality. The grounds for this position, though, are developed only in *S3*, to which we will return in Chapter Four.

Meanwhile, it should be noted that De Morgan's treatment of these two propositional forms, and of the inferences associated with them, is problematic in several ways. To begin with, it is odd that in the example which accompanies his analyses, equality is treated in two ways at once: first, as the predicate in "All right angles are equal," whose copula is "is," and then as part of the second copula "is equal to." What, if anything, is the relation between these two forms?

This question recurs with greater force when we examine De Morgan's proof of the Pythagorean Theorem (SDM, 221–23). Some of these problems may be due to the fact that several different kinds of equality are present at the same time –equality of angles, of lines, of triangles, and of parallelograms. But De Morgan's vacillation between equality as a predicate and equality as a relation is vividly illustrated. In Step IV, he treats equality as a predicate, when he argues,

(2.3) All sides of a square are equal.
AB and BE are sides of a square.

AB and BE are equal.

It should be noted that here a compound subject occurs essentially, since it must be stipulated both that AB and BE are sides of the same square, and also that they are equal to each other. This type of inference occurs several times in the proof. At Step VII, however, "is equal to" is treated as a copula, in the inference,

(2.4) BG is equal to BC.
BC is equal to BD.
―――――――――――
BG is equal to BD.

No attempt is made to reduce this inference to the common syllogism by the addition of a premise, as Hamilton would have demanded. Presumably it is just an instance of first figure, AAA.

Our confusion about the two uses of equality is compounded by the fact that in moving from Step IX to Step XII, De Morgan infers "EBAF is equal to BGHA" from "EBAF and BGHA are equal," without comment.

There are two other inferential steps which De Morgan takes in this proof which must give us pause. In one case he clearly uses multiple (double) subsumption. This is in the step,

(2.5) Two triangles having two angles of one equal to two angles of the other, and the interjacent sides equal, are equal in all respects.
BEG and BAC are two triangles having BEG and EBG, respectively, equal to BAC and ABC and the sides EB and BA equal.
―――――――――――
The triangles BEG and BAC are equal in all respects.

While this has a semi-syllogistic form, it is in fact much more complex than that. In addition to the use of multiple subsumption, additional clauses must be added, such as that side EB is the interjacent side for BEG and EBG. Whether the fully articulated inference could be construed as merely a case of multiple subsumption is an open question.

Finally, it should be noted that the minor premise of Step IV is inferred in a step which is never given, but which would have to be stated as,

(2.6) AB and BE are sides of ABEF.
ABEF is a square.
―――――――――――
AB and BE are sides of a square.

This is not a syllogism in De Morgan's sense and, in fact, its validation requires the use of a principle which invokes embedding terms ("ABEF" and "square") in a relational expression. We shall encounter a similar case when we discuss De Morgan's *Formal Logic*.

These problems suggest that the independence of logic and geometry, as we have called it, applies quite strikingly to De Morgan. Despite the novelty of his doctrine of the two copulas, he presents a fairly conservative account of geometrical reasoning. Unlike most authors, he did attempt to provide a quasi-syllogistic analysis of at least one geometrical proof. Yet, when pressed just a bit, this syllogistic demonstration breaks down at several key points where intuitive inferences must be substituted for the syllogism.

Returning to the analysis of propositions, we find a natural final objection: Why should there be only two copulas? If *is* and *is equal to* are both acceptable copulas, why not *is greater than*? It would, in fact, be quite peculiar to treat *is greater than* differently than *is equal to*, since they are essentially the same kind of relation. De Morgan says that any geometrical proposition can be expressed using one of his two copulas, yet he does not provide any systematic evidence that this is so. (The nature of the book in which this appears, of course, might dictate that this not be pursued further.) Even if this reduction can be carried out, what grounds are there for claiming that the reduced form is the perspicuous one? And, given the reduction, can we be assured that logical principles will allow us to handle all the essential inferences, using the reduced form? We should remember that Hamilton provided a general principle to support arguments involving the transitivity of equality, but that the logic which he had available did not actually allow the inference to be carried out.

We have raised the possibility of adding "is greater than" to the list of copulas. But why stop with it? It appears that any two-place relational expression would work equally well. If, as we have suggested, De Morgan had already adopted a relational analysis of the proposition, asserting that a proposition consists of a subject term related in some way to a predicate term, then it seems that any relation will do. We shall see that this is just the move which De Morgan makes in *Formal Logic*.

If rules to handle the transitively relational inferences for "is" and "is equal to" are legitimate and primitive, why not similar rules for all other transitive relatio"? Why isn't the transitivity of "is greater than" an equally good candidate, or even of "is tied to" and "causes"?

In this early work, though, De Morgan seems unwilling to take this extreme position, for he believes that the transitivity of "is greater than" can be given a purely syllogistic treatment.

Reasoning *a fortiori*, is that contained in Fig.I.1. in a different form, thus: A is greater than B, B is greater than C; *a fortiori* A is greater than C; which may be also stated as follows:

The whole of B is contained in A,
The whole of C is contained in B,
Therefore C is contained in A.

The premises of the second do not necessarily imply as much as those of the first; the complete reduction we leave to the student.(SDM, 220)

Since he comments no further on this topic, it is difficult to see just what he is claiming. However, at least a few points can be clarified. To begin with, Fig.I.1. is the standard first figure AAA of the syllogism which, however, is stated in a somewhat unusual way. De Morgan supposes that we are given three figures, a circle, a triangle, and a square, and we wish to assert propositions concerning which of them are contained in which of the others. Applying first figure AAA to them, we would argue:

(2.7) All the ○ is in the △.
All the □ is in the ○.
―――――――――――――
All the □ is in the △.

This formulation may appear to turn this inference into one involving class inclusion, or even the part-whole relation, especially since on an earlier page De Morgan uses "is contained in" rather than "is in." This would make it appear that we have yet another relation whose transitivity is taken as a primitive rule. However, his earlier discussion of this formulation makes it clear that this is not so, since to say, for instance, that "All the circle is in the triangle" is merely to say that "Every point of the circle is a point of the triangle." This means, presumably, that the syllogistic form of the *a fortiori* inference would be construed as,

(2.8) All the B is in the A.
All the C is in the B.
―――――――――――――
All the C is in the A.

Since there is no indication of the type of entities for which the letters stand, we cannot give an idiomatic translation into standard categorical form as we did with "All the circle is in the triangle." However, using "element" in a very general sense, it might be stated as:

(2.9) Every element of B is an element of A.
Every element of C is an element of B.
———————————————
Every element of C is an element of A.

This is obviously a traditional syllogism; unfortunately, though, it is not clear that it truly represents the original *a fortiori* inference, and for two reasons.

To begin with, as De Morgan himself notes, the premises of the syllogistic form (2.9) do not imply as much as those of the original form, as they are often construed. In many instances, the original argument suggests that A is greater than C by a margin greater than that by which A exceeds B. There is no suggestion of this, though, in either (2.8) or (2.9). In saying that "the complete reduction is left to the student," De Morgan may be suggesting that a more elaborate treatment can remedy this defect; but he does not tell us what it is. We shall soon see that he does give such an account in his first work on logic.

This issue is related to the more fundamental question of whether "A is greater than B" is really equivalent to "The whole of B is contained in A," since "greater than" is an asymmetric relation and "is contained in" is merely nonsymmetric. As De Morgan will show in his first logical work, something must be added to get a proposition equivalent to the original "A is greater than B." The result would be something like, "B is partially contained in A."

Even if we can translate "A is greater than B" into terms which are also asymmetric, we may still wonder whether the two statements are true equivalents. In many situations, A is greater than B just in case B is partially contained in A; for instance, if A is a line of which B is a part. But this need not always be the case. I may have two lines, or two rulers, one of which is greater than the other, but neither of which is in any sense contained in the other. Alternatively, the number 5 is greater than the number 3, but the number 3 is not "contained" in the number 5; the same may be said for five square miles and three square miles.

These difficulties of reducing even the transitivity part of *a fortiori* arguments to syllogistic form must make us wonder, once again, why De Morgan did not just add "is greater than" to his list of copulas, and take its transitivity as another instance of first figure AAA.

Before considering De Morgan's first work in logic, we should notice that in this chapter he makes several assertions which could be construed as embodying the germ of his later relational account of reasoning. He says at one point,

> It is evident that no assertion can be the direct and necessary consequence of two others, unless those two contain something in common, or which is spoken of in both.(SDM, 210)

Later, he says:

> In all that has gone before we may perceive that the validity of an argument depends upon two distinct considerations,–(1) the truth of the relations assumed, or represented to have been proved before; (2) the manner in which these facts are combined so as to produce new relations; in which last the *reasoning* properly consists. . . . We are accustomed to talk of mathematical *reasoning* as above all other, in point of accuracy and soundness. This, if by the term *reasoning* we mean the comparing together of different ideas and producing other ideas from the comparison, is not correct, for in this view mathematical reasonings and all other reasonings correspond exactly.(SDM, 228)

These passages suggest that reasoning consists merely in the generation of new relations from old ones, with no suggestion that there are only two preferred relations, *is* and *is equal to*. At the same time, the very notion that reasoning consists in "comparing" ideas may involve a restriction on the scope of the allowable relations and reasonings.

2. FIRST NOTIONS OF LOGIC

De Morgan's first work in logic was a pamphlet, "First Notions of Logic," which was published in 1839 and included as the first chapter of *Formal Logic* in 1847. It is intended as an introduction to the strictly Aristotelian portion of logic; as such, we need not think that it represents De Morgan's full thinking about logic even in 1839. Nevertheless, in presenting this system he makes some fundamental claims that are of some interest, and he also provides an interesting analysis of *a fortiori* arguments.

His conception of reasoning is similar to that which he held in 1831.

Every act of reasoning must mainly consist in comparing together different things, and either finding out, or recalling from previous knowledge, the points in which they resemble or differ from each other. That particular part of reasoning which is called *inference*, consists in the comparison of several and different things with one and the same other thing; and ascertaining the resemblances, or differences, of the several things, by means of the points in which they resemble, or differ from, the thing with which all are compared.

There must then be some propositions already obtained before any inference can be drawn. All propositions are either assertions or denials, and are thus divided into *affirmative* and *negative*. Thus, X is Y, and X is not Y, are the two forms to which all propositions may be reduced.(FL, 2)[4]

He notes that "it will frequently happen that much circumlocution is needed to reduce propositions" to these two forms. Thus, "If he should come tomorrow, he will probably stay till Monday" must be rewritten as "The happening of his arrival tomorrow is an event from which it may be inferred as probable that he will stay till Monday." Despite the artificiality of these circumlocutions, De Morgan strongly asserts that such a reduction is always possible.

After discussing the four traditional propositional forms, De Morgan considers the syllogism, that is, "the manner of making two assertions give a third." Oddly enough, he begins his discussion of the syllogism with an example which is not, as stated, a syllogism:

(2.10) X is the same colour as Y.
Z is the same colour as Y.
―――――――――――
X is the same colour as Z.

Also, there is,

(2.11) X is not the same colour as Y.
Z is the same colour as Y.
―――――――――――
X is not the same colour as Z.

He then notes that these examples contain:

the essential parts of all inference, which consists in comparing two things with a third, and finding from their agreement or difference with that third, their agreement or difference with one another. Thus, Every X is Y, every Z is Y, allows us to infer that X and Z have all those qualities in common which are necessary to Y. Again, from every X is Y, and "No Z is Y," we infer that X and Z differ from one another in all particulars which are necessary to Y. The preceding forms, however, though they represent common reasoning better than the ordinary syllogism, to which we are now coming, do not constitute the ultimate form of inference. Simple *identity* or *non-identity* is the ultimate state to which every assertion may be reduced; and we shall, therefore, first ask, from what identities, &c., can other identities, &c., be produced? Again, since we name objects in species, each species consisting of a number of individuals, and since our assertion may include all or only part of a species, it is further necessary to ask, in every instance, to what extent the conclusion drawn is true, whether of all, or only a part? (FL, 8–9)

De Morgan then uses identity to construe "Every living man respires" as "Every living man is one of the things which respire."

Having given this analysis of propositions and inference, De Morgan provides an account of the syllogism quite similar to that which he gave in 1831. He immediately asks us to imagine that all living men are enclosed in a triangle, and that all respiring objects are enclosed in a circle, so that this assertion is interpreted to mean that the triangle is contained in the circle. With this geometrical picture in mind, De Morgan finds that there are really six distinct propositional forms:

(2.12) All the △ is in the ○.
None of the △ is in the ○.
Some of the △ is in the ○.
Some of the △ is not in the ○.
All the ○ is in the △.
Some of the ○ is not in the △. (FL, 10)

De Morgan then constructs his syllogistic system in two stages: first, by considering the possible combinations of these forms and the conclusions which follow from them; and then, by interpreting all the traditional syllogistic forms in his system. He concludes his account by making a quite sweeping assertion of the adequacy of the system which he has obtained.

Every assertion which can be made upon two things by comparison with any third, that is, every simple inference, can be reduced to one of the preceding forms.(FL, 21)

To support this claim, De Morgan shows how to reduce four sample arguments to syllogistic form. Two of these are unproblematic, but two are not: the *a fortiori* argument and the hypothetical syllogism. He concludes the pamphlet with short comments on indirect demonstration, conversion, and a complex propositional inference.

Before turning to De Morgan's analysis of the *a fortiori* argument, it will be useful to examine the analysis of propositions and inferences which he presents. On the propositional side, De Morgan begins with the plausible comment that all propositions contain a sign of assertion or denial; and, presumably, there must be some linguistic sign to say which one it is. De Morgan adds to this the more controversial claim that this means that every proposition is of the form "X is Y" or "X is not Y." This may seem reasonable on the level of categorical propositions, but it seems very artificial to translate hypothetical propositions into categorical form in order to sustain this claim. Furthermore, one may well wonder just how an assertion is being made in saying, "If Mary is at work then John is at home," since neither "is" carries assertive force, and there is no other sign of assertion in the usual form.

However we resolve these difficulties, we must question any analysis of the proposition which demands that every proposition be converted into a form that contains "is." We seem headed for a Procrustean bed, especially when this standard form is taken to contain something much more specific as well. It appears that for De Morgan, "is" is not only a sign of assertion, but also a sign of identity, that "*simple identity* or *non-identity* is the ultimate state to which every assertion may be reduced." When the "ultimate" form of "every living man respires" is "every living man is one of the things which respire," the verb "respires" has become "is one of the things which respire"; and what is really asserted is that "every living man is identical with one of the things which respire"(FL, 9). Why shouldn't one think of this as the "ultimate" form? It must be admitted that

$$(x)(Lx \supset Rx)$$

is logically equivalent to

$$(x)(Lx \supset (\exists y)(x = y \ \& \ Ry)),$$

but why is the latter more basic than the first? We shall soon see the role which this claim plays in a series of formal developments in De Morgan's logic; but for the moment, we need only note that it is a strong and unsupported thesis. To make good this thesis, De Morgan must not only show that for every proposition there is an equivalent which is expressed in terms of identity or non-identity; he must also show that the identity (or non-identity) form is, in some sense, the more basic.

The other general point to notice is the abstract way in which De Morgan characterizes inference. Aristotelian logicians had often assumed a traditional analysis of categorical propositions, and then said that inference involved finding a middle term to link two such propositions in order to generate a valid conclusion. De Morgan states a more general form of this claim, by saying that inference "consists in the comparison of several and different things with one and the same other thing"(FL, 2); and later he remarks on "the essential parts of all inference, which consists in comparing two things with a third, and finding from their agreement or difference with that third, their agreement or difference with one another"(FL, 9). This is a quite general account of reasoning, because it contains no insistence that these comparisons involve categorical propositions or the Aristotelian laws. It is striking that De Morgan takes the non-Aristotelian (2.10) as the primary instance of this. His claim is not one of unlimited generality, though, because it uses the notion of comparison, of noting similarities and differences between two items and some other "middle" item. The point may be put by saying that a wider generalization would involve somehow relating two terms by means of a third term, but that De Morgan limits this to relations involving similarity or difference.

We may conclude that De Morgan's characterization of inference is not inherently syllogistic, and that it only becomes so when a very specific mode of propositional analysis is introduced. This means that inferences such as (2.10) must eventually be reduced to syllogistic form. This is, of course, just another case of the transitivity of a relation; and, though De Morgan does not say how (2.10) is to be reduced, he does have an extensive discussion of the *a fortiori* argument, to which we now turn (FL, 22–25).

This time he considers the argument,

(2.13) X is less than Y.
Y is less than Z.
―――――――――
X is less than Z.

He notes that this is not a syllogism, since it is really of the form,

(2.14) X is (a magnitude less than Y).
Y is (a magnitude less than Z).
―――――――――――――――
X is (a magnitude less than Z).

which lacks a middle term. In order to reduce this to syllogistic form, we must add a premise just as we did earlier with (1.18). De Morgan suggests,

(2.15) If Y is a magnitude less than Z, then every magnitude less than Y is less than Z.
Y is a magnitude less than Z.

Every magnitude less than Y is a magnitude less than Z.
X is a magnitude less than Y.

X is a magnitude less than Z.

This reconstruction suffers from requiring a first premise which is absolutely specific to X, Y, and Z, while an adequate reconstruction should utilize a general principle. Multiple subsumption would be required for this; and if this is allowed, one might just as well use the analogue to (1.24):

(2.16) If one thing is less than a second thing, and the second thing is less than a third thing, then the first thing is less than the third thing.
X is less than Y and Y is less than Z.

X is less than Z.

De Morgan, on the other hand, rejects the strategy of (2.15) on the ground that its additional major premise "will be found to require precisely the same considerations repeated; for the original inference was nothing more"(FL, 23). It is difficult to be sure of the exact meaning of this cryptic remark, but on the most obvious interpretation it faces a serious objection. De Morgan may well be saying that (2.15) is defective since its major premise just embodies the rule behind (2.13); and thus, adding this premise means trivializing the claim that the argument can be put into syllogistic form. A traditionalist like Mansel would argue that this is perfectly legitimate when the principle involved is a material principle rather than a formal principle; and the transitivity of "less than" seems to be a material principle. De Morgan will have a great deal to say about the distinction between the formal and the material later on, but for the moment we must leave the issue in limbo.

In fact, De Morgan claims that (2.13) "involves more than any simple syllogism can express."

When we say that X is less than Y, we say that if X were applied to Y, every part of X would match a part of Y, and there would be parts of Y remaining over. But when we say, "Every X is Y," meaning the premise of a common syllogism, we say that every instance of X is an instance of Y, without saying any thing as to whether there are or are not instances of Y still left, after those which are also X are taken away.(FL, 23)

An ordinary syllogism "which shall correspond with" (2.13) would have to be written as,

(2.17) Every X is Y; (Some Ys are not Xs).
Every Y is Z; (Some Zs are not Ys).
———————————————
Every X is Z; (Some Zs are not Xs).

If each premise is viewed as a conjunction, the conjunctive conclusion follows from the premises. In fact, the right-hand conjunct of either premise (but not both) could be dropped and the syllogism remain valid. The remainder of De Morgan's discussion of this argument, though, can only be said to be confusing. Such an argument as (2.13) is called an *a fortiori* argument because, as we have seen in (2.17), "the premises are more than sufficient to prove the conclusions and the extent of the conclusion is thereby greater than its mere form would indicate." When "*a fortiori*" is included in the conclusion of (2.13), the whole argument means that:

the extent to which X is less than Z must be greater than that to which X is less than Y, or Y than Z. In the syllogism last written [2.17], either of the bracketted premises might be struck out without destroying the conclusion; which last would, however, be weakened. As it stands, then the part of the conclusion, "Some Zs are not Xs," follows *a fortiori*.(FL, 24)

From this discussion it would appear that De Morgan is contending that *a fortiori* arguments are of the form (2.17). However, he concludes somewhat confusedly by saying that "The argument *a fortiori* may then be defined as a universally affirmative syllogism, in which both of the premises are shown to be less than the whole truth, or greater"(FL, 24). The full *a fortiori* form would then be stated as,

(2.18) All the Xs make up part (and part only) of the Ys.
All the Ys make up part (and part only) of the Zs.
———————————————
All the Xs make up *part of part* (only) of the Zs.

where "the words in italics mark that quality of the conclusion from which the argument is called *a fortiori*"(FL, 24).

This new form presents many problems. Its syllogistic status is as cloudy as the original, and we might do well to remain with (2.17). In writing of "part of part," De Morgan has introduced something like the composition of relations, which he will later claim is a general phenomenon of which the syllogism is but an instance, rather than something which can be reduced to the syllogism.

Furthermore, the conclusion of (2.18) does not capture his intention of including the fact that the excess of Z over X is greater than the excess of Z over Y and of Y over X. This is due to the simple fact that the conclusion of (2.18) does not refer to Y at all. It says only that there is something which is such that X is a part (only) of that something, and that that something is a part (only) of Z. It does not say that this is Y.

A full syllogistic representation of the *a fortiori* argument must exhibit three separate features:

a. "Is less than" is transitive.
b. "Is less than" is asymmetric.
c. The amount by which X is less than Z is greater than the amount by which X is less than Y and the amount by which Y is less than Z.

Using only the left-hand conjuncts of (2.17) will capture transitivity; adding the right-hand conjuncts, or moving on to (2.18), would include asymmetry as well. But none of these formulations can include feature *c*, which requires a second level of comparisons of greater and less. This is a feature of such complexity that it is doubtful that it could be formulated without a very substantial increase in notational power.

Leaving aside this last feature, we may ask whether either (2.17) or (2.18) is an adequate formal representation of (2.13). We have already remarked that (2.18) seems not to be syllogistic. In any case, both (2.17) and (2.18) involve translating comparisons of magnitude into class inclusion and exclusion statements, and we may raise the same questions about this as we did in the first section. De Morgan writes of constructing a syllogism which will "correspond" to (2.13); and in one sense (2.17) and (2.18) do correspond, in that they both involve a transitive and asymmetric relation. Furthermore, there are some cases where the magnitude of X is less than that of Y because X is a proper subset of Y–e.g., when Y is a line of which X is a part. But, this is surely not the case with all magnitudes; or at the very least, an analysis of magnitude statements would have to be given which would have this feature.

One cannot help but be puzzled by the fact that De Morgan tried so hard to reduce *a fortiori* arguments to syllogistic terms. Our discussion of *SDM* suggested that once two copulas are admitted, with equally basic formal laws, there is no reason not to admit additional ones. De Morgan could have taken the transitivity of *is less than* as a basic rule, just as he did that of *is* and *is equal to* in *SDM*. In *FN*, though, De Morgan has admitted only one affirmative copula, rather than the two copulas of *SDM*. *FL* will take the dramatically different tack of allowing an indefinitely large number of copulas.

3. RELATIONS AND IDENTITY

From 1833 to 1843, De Morgan was a prolific contributor to the *Penny Cyclopaedia*, published by the Society for the Diffusion of Useful Knowledge, of which he was an active member. He wrote most of the articles on mathematics for the encyclopaedia, a large number on astronomy as well, along with biographies of prominent mathematicians and scientists. It is estimated that he wrote about 850 articles, a substantial portion of this encyclopedia of twenty-seven large double-columned volumes. This contribution is immense, especially when De Morgan's other writings of this period are considered.[5]

De Morgan did not write the main article on logic, entitled "Organon." He did, however, write the article "Syllogism" and about thirty short articles on logical topics. For our purposes, we need only consider his articles "Relation (Mathematical)" and "Syllogism."

In both of these articles, he adopts a very conservative stance on the adequacy of the syllogism, while embedding his account in a relational analysis which promises more to come. He begins his article on relations by saying:

> All reasoning is the discovery of relations which are not evident from those which are; or rather, since the proposed result is sometimes evident in itself, reasoning is the establishment of one relation as a necessary consequence of others. The term relation would be difficult to define in a manner satisfactory to all; it is enough for our present purpose to say a relation exists between any two objects, whether of sense or intellect, whenever they have anything in common; that is to say, the common point, whatever it

may be, may be made the means of referring one to the other, or bringing our thoughts from one to the other, so as to think of both at the same time, and to compare the two. All the manifold senses of the word may be derived from this one: the relationship of blood implies a common ancestry; the relationship of office, common duties. In mathematics the relation of greater, equal, or less, implies that one of the magnitudes is the same as to quantity with part or all of the other, and so on. Sameness in every respect would constitute identity; sameness in one or more respects, relation.[6]

Since this is an article on mathematical relation, De Morgan devotes much of it to considering the relations which occur in algebra, but which are all signified by the expression " = ," *is equal to*. "In algebra there is a great want of terms, as well as of signs, significative of relation; we have in fact only one sign, = , and one phrase for it, *is equal to*. Hence proceeds either want of distinction between different relations, or the want of distinctness which arises out of such a generalization of the sign as will bring all its meanings under one." After discussing three algebraic relations, he says:

It might perhaps be said that though we have constantly used the word relation, yet we have considered nothing but identity, that is, either identity of magnitude, form, process, or properties; but that term in common life refers to something short of complete identity, frequently meaning mere connection, and sometimes only analogy, or even nothing more than resemblance. We answer, that relation always refers to identity of some sort. For example, there is a relation between the position of the sun and moon and the state of the ocean. Here the word means merely a connection; but this connection involves an absolute identity: having given the position of those heavenly bodies with respect to any place, together with the direction and quantity of their motions, the height of the water at that place is connected with the quantities which express those positions and motions by an equation or a mathematical identity. Resemblance again means identity in some respect, or near approach of identity: analogy, a term generally applied to relations of similarity, will be found to admit of the word sameness being used instead of similarity. Thus, when we say that substance is formed from *substare* in a manner similar to that in which distance is formed from *distare*, the analogy asserted is one of absolute identity (of mode of derivation).[7]

Once again, De Morgan begins with what sounds like a highly general account of reasoning, thought of as establishing one relation on the basis of others. However, the generality of this account is immediately reduced drastically by what one might call the Identity Theory of Relations. He seems to be asserting in both passages that relations not only involve identities, but that in some sense they are reducible to identities. Thus, in the first quoted passage, he says that a relation exists between two objects whenever they have anything in common; this common point may be "made the means of referring one to the other, so as to think of both at the same time, and to compare the two." De Morgan here seems to be echoing his claim in *FN* that all reasoning involves comparison between two items by means of a third; and how could two items be compared, if they did not have anything in common? He then provides what seem to be reductions of relational statements to identity statements. Two of his three examples are:

> X is a blood relation of Y = $(\exists W)(\exists Z)$(W = Z & W is an ancestor of X & Z is an ancestor of Y).
>
> X is greater than Y = $(\exists W)$(W is a part (only) of X & W is the same as to quantity as Y).

The second quotation makes a similar point. De Morgan refuses to apologize for his emphasis upon identity in an article on relation, by saying that even when a relation is merely one of connection, or of analogy, or of resemblance, "relation always refers to identity of some sort." In the simplest cases,

> X resembles Y = $(\exists P)(\exists Q)$(P = Q & P(X) & Q(Y)).
>
> A is formed from B in a manner similar to that in which C is formed from D = $(\exists M)(\exists N)$(M is a mode of derivation & A is formed from B by M & N is a mode of derivation & C is formed from D by N & M = N).

It should be noticed, to begin with, that his analysis will work only if we are allowed to quantify over a wide range of entities–individuals, properties, modes of derivation, and probably many other types of entities as well. It should be noted also that in each case, the translation has been formulated so as to include an identity relation explicitly, even though this may seem rather long-winded. The only exception is in the analysis of "greater than," where this has been replaced by "the same as to quantity." Even here, an identity could be created:

> X is greater than Y = $(\exists W)(\exists Q)(\exists R)$(W is a proper part of X & Q is the quantity of W & R is the quantity of Y & W = R).

I have interpreted De Morgan's claims here in the strongest possible way: that all relational statements can be reduced to identity statements. He seems not to be asserting only that identity statements are implied by relational statements; he seems to be claiming, rather, that identity is the only relation we need. Standing in a relation means having something in common, the nature of which is specified for the particular relation in question; in his words, "relation always refers to identity of some sort." It would seem that limiting the scope of his discussion to identity relations would not be justified unless all relations could somehow be reduced to the identity relation.

Unfortunately, a glance at the proposed translations shows that this is not correct, for in almost all cases other relations occur in the analysis. This includes the relations *being an ancestor of*, *being a proper part of*, and even the three-place relation *forming by means of*. Only resemblance escapes this problem, and even here this may be because the analysis is too weak. Resemblance in ordinary parlance involves more than just sharing a property, and an attempt to specify this might throw us back into another relation.

It is ironic that De Morgan, the father of the logic of relations, should have flirted with the idea that there is only one relation, identity. The greatly increased range of *FL* will show us that he came to abandon this view.

De Morgan begins his article "Syllogism" with the strong claim,

> All that is called reasoning, and which cannot be made syllogistic, is not reasoning at all; and all which cannot easily be made syllogistic, is obscure: for the syllogism is the simple form in which the act of reasoning is an intuition.[8]

This has little definite meaning, of course, in the absence of a full characterization of what a syllogism is. "Syllogism" has sometimes meant any putatively deductive argument, while sometimes it has been confined to the Aristotelian variety. De Morgan's initial characterization of a syllogism is similarly broad–"Every sentence in which the conclusion is a necessary consequence of previous assertions contained in that same sentence, is a syllogism, provided that the conclusion be obtained from two distinct assertions, and two only."[9] The restriction to arguments with two premises does involve the substantive claim that all valid arguments can be structured so as to include only two premises. Of course, if a valid argument contained more than two premises, their number could be reduced to two by conjoining some of them. Thus, his assertion has content only if he can specify the form of allowable assertions, which he does in the next sentence by saying that "every assertion may be reduced to one of four forms"–that is, the traditional categorical forms. He then concludes, "Since every conclusion must be drawn from the comparison of two things with a third, a syllogism consists of two propositions, in each of which the same term occurs compared with another." The bulk of the article is then devoted to stating and illustrating the syllogistic forms, and to discussing, from a syllogistic standpoint, types of arguments which require some ingenuity to be made syllogistic. In this latter group he does not include the transitivity and *a fortiori* arguments with which we have been concerned. All of this is done in a very straightforward manner, with none of the novelties which he introduced in *FN*. This form was probably dictated by the place of this article in an encyclopedia. At the same time, it seems clear that he is endorsing the traditional claims to the completeness of the syllogistic system, and not simply reporting that the friends of the syllogism would do so.

In this chapter, we have seen De Morgan's ambivalence toward traditional syllogistic analyses. In the next chapter, we will confront his dramatic innovations.

GENERALIZING THE COPULA

During the period from 1846 to 1850, De Morgan's thinking about relations developed in two ways which were essential for motivating a formal logic of relations. The first was the recognition that relations are a subject for logical study in their own right. The form which this recognition took in De Morgan's work was unusual, and it gave rise to philosophical problems; but it was important in shaping his entire approach to the logic of relations.

We have already seen that in 1831 De Morgan was willing to countenance two copulas, *is* and *is equal to*. These copulas are formally similar in that the laws of the syllogism apply to each. In his very important *Formal Logic* (1847), De Morgan recognized that there is no reason to limit logic to these two copulas. Any relational terms will satisfy the syllogistic laws, as long as the "formal" properties of transitivity and convertibility apply to them. The relation *is tied to* is as good as *is equal to*. One can even study the syllogistic laws which follow merely from transitivity, and then see the effect of adding convertibility. Thus was born the *doctrine of the abstract copula*. Logic is conceived of as dealing with the consequences of the formal properties which relations possess. Relations had become the subject matter of logic.

De Morgan's second crucial foundation for the logic of relations is found in his memoir "On the Syllogism, II" of 1850. The logic of relations plays a much more central role here than it did in *FL*, and the result is a very significant generalization of that account. There is no reason why logic should be limited to relations which have certain formal properties. For instance, to say that equality is transitive is to say that the following argument is valid:

A is equal to B.
B is equal to C.
―――――――――
A is equal to C.

In this argument, one infers a relation between A and C, from one which holds between A and B, and one which holds between B and C. In this case, all three relations are the same; but they need not be. One could argue, for instance,

A is a brother of B.
B is a parent of C.
―――――――――
A is an uncle of C.

This is the basis for De Morgan's theory of the *bicopular syllogism* which embodies the *composition of relations*. In the composition of relations, De Morgan has found the essential notion for a formal logic of relations. He then uses both the doctrine of the abstract copula and the composition of relations to serve as the foundation for his theory of syllogistic laws.

In *FL* and *S2*, De Morgan also expresses an interest in a type of inference which involves the composition of a relation with a class, rather than the composition of a relation with a relation. These inferences are "oblique" inferences, such as "Every man is an animal; therefore, every head of a man is the head of an animal." De Morgan develops a generalized version of the traditional *dictum de omni et nullo* to handle this and many other cases. We will see that the need to handle oblique inferences makes De Morgan's version much more radical than it might appear and that this very feature creates severe problems of application.

1. THE ABSTRACT COPULA

De Morgan's serious work in logic began in 1846, with the preparation of the first of a series of five papers on logic which appeared from 1847 to 1863 in the *Transactions of the Cambridge Philosophical Society*.[1] This series, often referred to as "On the Syllogism," contains a bewildering variety of material: new systems of notation, generalizations of the syllogism, philosophy of logic, attacks on Sir William Hamilton and other opponents, and historical comments. For our purposes, the first paper in this series does not require much comment, for its main purpose is merely to develop a new approach and notation for the traditional syllogism. The result is that it discusses little that is relevant to the logic of relations. It should be noted, though, that it contains the same endorsement of the sufficiency of traditional logic which we found in "First Notions."

After providing a new notation in *S1* for the four types of categorical propositions, De Morgan asserts the sufficiency of these forms by saying that "All expressions that have any meaning can of course be reduced to one of these forms." The resulting analysis of inference is similar to that which we have noted in the *Penny Cyclopaedia*:

> Inference must arise from bringing each two things which are to be compared into comparison with a third. Many comparisons may be made at once, but there must be this process in every one. When the comparison is that of identity, of *is* or *is not*, it can only be, in its ultimate or individual case, one of the two following;–'This X is a Y, this Z is the very same Y, therefore this X is this Z;' or else 'This X is a Y, this Z is not the very same Y, therefore this X is not this Z.' And collectively, it must be either 'Each of these Xs is a Y; each of these Ys is a Z; therefore each of these Xs is a Z;' or else 'Each of these Xs is a Y, no one of these Ys is a Z, therefore no one of these Xs is a Z.'
>
> All that is essential then to a syllogism is that its premises shall mention a number of Ys, of each of which they shall affirm either that it is both X and Z, or that it is one and not the other. (S1, 8)

Inference, that is, involves comparing two objects through a third object. This may be construed as a limitation on a fully relational account: such an account would hold that inference requires relating two objects through a third object; but it is not clear that all relations involve comparison. De Morgan's analysis embodies a further restriction as well, for it states that this comparison is ultimately that of identity or non-identity. We will see that the identity relation will continue to exert a strong force on De Morgan's analysis of the copula.

The syllogistic system of *S1* was incorporated into *Formal Logic*, which was published in November of 1847.[2] This innovative work was the source of much of De Morgan's reputation as a logician. In its Preface, De Morgan listed its eight major innovations. This included the heavy use of contraries; a theory of complex propositions; a new system of syllogistic forms; avoidance of syllogistic figure; a means of dealing with compound names which is, in reality, a logic of classes; the numerically definite syllogism; and the replacement of the "old doctrine of modals" by the theory of probability. The innovation in which we shall be most interested, however, is the first one which he lists.

> In the form of the proposition, the copula is made as abstract as the terms: or is considered as obeying only those conditions which are necessary to inference. (FL, ix)

In this section, we shall attempt to understand this innovation.

In Chapter III, "On the Abstract Form of the Proposition," De Morgan discusses both the terms and the copulas of propositions. He points out that the relations between words, ideas and objects are notoriously controversial, and that every attempt must be made to develop the laws of logic so as not to presuppose any one theory. We want to abstract from these competing theories "to arrive at a distinct abstraction of so much of the idea we are concerned with, as is itself the precedent reason, if it be right so to speak, of the law in question"(FL, 54). In order to do this, De Morgan takes an abstract view not only of the terms of propositions, but of the copula as well.

Aristotle and others had used letters to stand for terms in simple propositions. In order to do logic, we need not decide whether they are to stand for names, ideas, or things. We need only settle upon a consistent interpretation to use throughout our discussions. Thus, only letters need be used to represent terms, without presupposing some particular method of analysis.

The same strategy may be used for the copula. In fact, as De Morgan notes, though the task of formal logic is to deal with "is" and "is not," the full meaning of the copula will depend upon that given to the terms; and, conversely, our treatment of the copula should not prejudice our interpretation of the terms. How, then, might the statement "man is animal" be interpreted? There are at least three senses of "is," depending upon the objects being talked about.

> Speak of *names*, and say "man *is* animal": the *is* is here an *is* of applicability; to whatsoever (idea, object, &c.) man is a name to be applied, to that same (idea, object, &c.) animal is a name to be applied. As to ideas, the *is* is an *is* of possession of all essential characteristics; *man* is an idea which possesses, contains, presents, all that is constitutive of the idea *animal*. As to absolute external objects, the *is* is an *is* of identity, the most common and positive use of the word. Every man *is* one of the animals; touch him, you touch an animal, destroy him, you destroy an animal. (FL, 57)

If we wish to abstract not only from the meanings of the terms, but also from that of the copulas, we must ask, what properties do these three senses share? De Morgan's answer is that what they share is not some common meaning, but rather "common laws of inference." To see what properties of "is" are required for inference, we will consider the specific case of the "doubly singular" proposition, "this one A is this one B." Three laws of inference govern the "is" in this proposition:

1. It is indifferent to conversion; that is, "A is B" and "B is A" have the same meaning. It is *convertible*.
2. If the connexion exists between one term and each of two others, it must exist between the two others. It is *transitive*.
3. *Is* and *is not* are contradictory alternatives; that is, exactly one of A is B and A is not B must be true (FL, 57–58).

Several pages later, De Morgan provides a more exhaustive treatment of five main senses of "is." The *first* sense is that of *absolute identity*, from which the three conditions above were derived. As we might expect, this sense is the primary one for De Morgan, because he thinks that propositions containing other senses can be reduced to absolute identity. The *second* sense is *agreement in a particular or particulars*, as in "He is a negro in colour." This can be reduced to "He is a thing having the colour of one of the negroes." In general, "Every A is B in colour" becomes "Every A is a thing having the colour of one of the Bs." Interestingly enough, though, such propositional paraphrases only yield correspondingly valid syllogistic inferences in some cases (FL, 61–62). For instance, there is,

(3.1) No X is Y (in color).
Every Z is Y (in color).
―――――――――――――
No Z is X (in color).

from which it follows that,

No Z is X.

This can be brought into a syllogistic form, using the *is* of identity, by writing it as:

(3.2) No X is a thing having the color of one of the Ys.
Every Z is a thing having the color of one of the Ys.
―――――――――――――
No Z is X.

However, if we try to use the same strategy with,

(3.3) Some Ys are not Xs in color.
Every Y is Z in color.
―――――――――――――
Some Zs are not Xs in color.
―――――――――――――
Some Zs are not Xs.

we would obtain,

(3.4) Some Ys are not things having the color of one of the Xs.
Every Y is a thing having the color of one of the Zs.

Some Zs are not Xs.

By treating this as a traditional syllogism, we could legitimately conclude,"Some things having the color of one of the Zs are not things having the color of one of the Xs." But we clearly cannot deduce "Some Zs are not Xs" because the major and minor terms in the premises are the more complex terms. And all this holds despite the fact that "Some Zs are not Xs" does follow from the more complex syllogistic conclusion.

The *third* use of "is" expresses *possession of a quality*, as in "The rose is red." It also includes the closely related case of the connection of an idea with one of its constituent or associated ideas, as in "Prudence is desirable." According to De Morgan, this sense is not admitted into logic since, among other things, the logical conditions are not satisfied. He points out that "the rose is red" cannot be converted to "red is the rose," which is a mere "poetical inversion of the first assertion" and not logically true"(FL, 60). He might well have claimed that "red is the rose" is not even well formed, other than as a "poetical inversion," since its subject is not a substantive. For this reason, transitivity cannot even be formulated for this sense of "is," since it requires an adjective to do double duty as the predicate of one sentence and the subject of another. The result is that "the rose is red" is normally translated by the logician into "the rose is a red object," in which the "is" is both convertible and transitive. This convertible and transitive sense is really identity, for what is really asserted, on this analysis, would be more fully expressed as "the rose is identical with one of the red objects."

This same strategy applies to the *fourth* case, that of *species and genus*. We have already noted the three ways in which De Morgan would construe one such proposition, "Man is an animal":

Every object to which "man" is applied is one of the objects to which "animal" is applied.

Man is one of the ideas which possesses all that is constitutive of the idea *animal*.

Every man is one of the animals.

In each of these sentences, the "is" is the "is" of identity. The first and third statements are quantified identity statements; the second, a singular identity statement.

The *fifth* use of "is" is that found in *existence statements*. This use is independent of the other four uses, and De Morgan makes no further comments on it.

By means of these paraphrases, De Morgan has reduced all the uses of "is" (except that of existence) to the "is" of identity. He reads all such statements as either doubly singular identity statements,

This one A is this one B,

singly singular identity statements,

This one A is one of the Bs,

or quantified identity statements,

Every A is one of the Bs,

Some A is not one of the Bs.

In this way, the uses of "is" which do not satisfy the three formal conditions on the "is" of identity are reduced to uses which do.

De Morgan's apparent fixation on the "is" of identity is not the most novel feature of his treatment of the copula. In fact, he says, this formal treatment of the copula need not be limited to the traditional "is," in any of its senses.

> Every connection which can be invented and signified by the terms *is* and *is not*, so as to satisfy these three conditions, makes all the rules of logic true. No doubt absolute identity was the suggesting connexion from which all the others arose: just as arithmetic was the medium in which the forms and laws of algebra were suggested. But, as we now *invent algebras* by abstracting the forms and laws of operation, and fitting new meanings to them, so we have power to invent new meanings for all the forms of inference, in every way in which we have power to make meanings of *is* and *is not* which satisfy the above conditions. (FL, 58)

For instance, if "is" is allowed to mean "is tied to,"

(3.5) Every X is Y.
Some Zs are not Ys.

Some Zs are not Xs.

becomes

(3.6) Every X is tied to a Y.
Some Zs are not tied to Ys.

Some Zs are not tied to Xs.

In the broad sense of "is tied to," (3.6) is just as valid as (3.5).

De Morgan recognizes that not all inferences require that all three principles be satisfied, so that, for instance, first figure AAA requires only transitivity. Under this principle he considers arguments involving "greater than," "less than," and "is equal to" which are "allowed to pass for syllogisms, in the ordinary definition of that word"(FL, 59). He points out that the transitivity of "greater than" allows,

(1.18) A is greater than B.
B is greater than C.

A is greater than C.,

but that "greater than" does not satisfy those forms which depend upon the convertibility of the copula. The same is true of "less than."

> But *is* in the sense "is equal to" *does* satisfy all the conditions. This sense of *is*, namely agreement in magnitude, is the copula of the mathematician's syllogism, when he is reasoning on quantity only. (FL, 59)

It is not clear, incidentally, just what to make of De Morgan's comments about the mathematician's "is" of equality. On the one hand, the phrase "agreement in magnitude" suggests that he intended to include this case under his second common meaning of "is." On this reading, "A is equal to B (in length)" is "A is B (in length)," which could be paraphrased as "A is a thing having the length of B." Or, one might even say, "the length of A is the length of B." This would have the theoretical elegance of construing yet another copula in terms of identity. Yet, this seems counterintuitive in terms of the larger enterprise. We remember that De Morgan proposed two copulas in *SDM*, *is* and *is equal to*. He is now greatly expanding the list of copulas to include many other relational terms, and it would seem odd now to remove "is equal to" from the list.

Furthermore, it seems best to treat "is equal to" in the same way as "is greater than." After all, "A is greater than B" could be reduced to "A is one of the things which is greater than B." Why make an analogous reduction for "equal to" and not do so for "greater than"? The formal reason for not doing so with "greater than" may well be that, for reasons which we have seen in Chapter One, the transitivity of "greater than" cannot be reduced to that of the "is" of identity by syllogistic means. But we also noted in that chapter that exactly the same problem afflicts the transitivity of "equal to."

However this point may be resolved, De Morgan's fundamental claim is that the traditional syllogism is based on three formal properties of "is," and that all of the syllogistic moods and figures apply equally to any other copulas with these same formal properties. We will call this thesis the *doctrine of the abstract copula*. De Morgan realizes that some relations have these properties and others do not. Later in the book, and in another connection, he points out that "gives," in the sense of causal production, can be used in transitive inferences. The same is true of "brings," as in "murder brings death to the perpetrator" (FL, 312). Transitive inferences are sound in those cases in which an action or state is communicated, as in "Snow brings floods, floods bring destruction; therefore, snow brings destruction." Other kinds of relations, such as "father of," do not communicate in this way.

De Morgan's formal account of the syllogism in *FL* is not seriously influenced by the doctrine of the abstract copula; in this respect, *FL* is profoundly different from his later works. In *FL*, the intricate theory of the syllogism is developed solely in terms of the traditional copula. Nevertheless, the doctrine of the abstract copula does give a certain flavor to this account; for De Morgan is suggesting that the validity of syllogisms involving quantified syllogistic statements (such as "Every A is B") is grounded in the formal properties of identity in doubly singular identity statements ("This one A is this one B"). Though this reduction is not carried out explicitly in *FL*, it is done in De Morgan's later work, using primitive versions of instantiation and generalization (S, 160). For transitivity, consider first figure AAA:

(3.7) Every Y is Z.
Every X is Y.
―――――――――
Every X is Z.

The transitivity of identity establishes this form as follows: consider an arbitrary X; call it "a." Then *a* is identical with some Y, call it "b." By the first premise, *b* is identical with some Z, call it "c." So, *a* is identical with *b* and *b* is identical with *c*; therefore, by the transitivity of identity, *a* is identical with *c*. But this means that *a* is identical with some Z. Since *a* is an arbitrary X, this means that every X is Z.

To see how convertibility combines with transitivity, we can use,

(3.8) Some Xs are Ys.
Every Y is Z.
____
Some Zs are Xs.

Consider one of the Xs which is identical to a Y; call it "a." Then *a* is identical to some Y, call it "b." But then *b* is identical with some Z, call it "c." So, *a* is identical with *b* and *b* is identical with *c*; therefore, *a* is identical with *c*. By convertibility, *c* is identical with *a*. Since *c* is a Z and *a* is an X, some Z is an X.

De Morgan's general account of propositions and inferences is of the relational sort which we have already encountered. Propositions assert agreement or disagreement, more or less, between two names; and a syllogism is "the inference of the relation between two names from the relation of each of those names to a third"(FL, 88). Yet, there is an interesting shift of perspective. We have so far described "Every X is Y" as a statement that every X stands in a certain relation–the identity relation–to one of the Ys. But we may equally well see this form as stating a relationship between the two classes X and Y themselves. De Morgan's account of this is quite complicated, because he deals both with simple propositions (which include the traditional Aristotelian forms) and complex propositions (forms involving the full specification of a state of affairs, such as "All X are Y and some Y are not X"). Furthermore, he develops a full theory of the syllogism for both types of propositions. Fortunately, we can understand his approach by dealing only with the traditional simple case.

In De Morgan's analysis in *FL*, there are eight propositional forms, each of which can be interpreted as expressing a relation between the class of Xs and the class of Ys (FL, 72, 86, 336–38). These are:

Every X is Y.	X is a species of Y.
Some Xs are not Ys.	X is a non-species of Y.
Every Y is X.	X is a genus of Y.
Some Ys are not Xs.	X is a non-genus of Y.
No X is Y.	X is an external of Y.
Some Xs are Ys.	X is a non-external of Y.
No non-X is non-Y.	X is a complement of Y.
Some non-Xs are non-Ys.	X is a non-complement of Y.

In the simple syllogism, propositions are transformed so that the argument is of the form: X bears one of these relations to Y, Y bears one of these relations to Z; therefore, X bears one of these relations to Z. From this perspective,

(3.9) Some Xs are Ys.
No Z is Y.
―――――――――
Some Xs are not Zs.

contains the terms of its second premise in the wrong order. It may be rewritten, though, to read,

(3.10) Some Xs are Ys.
No Y is Z.
―――――――――
Some Xs are not Zs.

Each of the propositions in this argument expresses a relation between classes, so that the syllogism can be expressed by,

(3.11) X is a non-external of Y.
Y is an external of Z.
―――――――――
X is a non-species of Z.

The principle which governs this inference is: *Non-external of external is non-species*: or, if X is a non-external of something which is an external of Z, then X is a non-species of Z. This principle is not too dissimilar from the principle that if X is greater than something which is greater than Z, then X is greater than Z. The only difference is that in the latter case there is one relation, while in the former case there are three relations. In both instances there is something which might be called the *product* of two relations; it is just that in one case the relations are different and in the other they are the same. In other syllogistic cases, of course, just one relation may be involved, so that the principle behind first figure AAA is *species* of *species* is *species*. De Morgan will explore the notion of the product of two relations much more fully in *S2*.

De Morgan's views concerning the adequacy of the syllogism must have changed greatly between *S1* and *FL*, for *FL* contains a number of argument forms which he claims cannot be validated syllogistically.

This includes inferences utilizing a form of the algebra of classes; the syllogism of transposed quantity, which requires the notion of a one-one correspondence; and the following inference, which Hamilton also discussed:

(3.12) Most A are B.
Most A are C.
―――――――――――
Some B are C.

This is a simple instance of what De Morgan called the numerically definite syllogism, an attempt to introduce numerical quantifiers into logic (FL, 159, 187, 379).

More importantly for our purposes, De Morgan discusses two forms of relational non-syllogistic arguments. The first is a type of inference, the *oblique inference*, which had been mentioned by some earlier logicians such as Aristotle and Jungius (FL, 131–32). De Morgan's example of this type of inference is,

(3.13) Man is animal.
―――――――――――
The head of a man is the head of an animal.

According to De Morgan, transformations of this type are often needed in the formation of the premises of a syllogism. That is, the conclusion of (3.13) may be used in a syllogism, but the inference is neither a syllogistic inference nor reducible to one. This form of inference is so important that we will devote all of Section Three to it.

De Morgan also has a brief discussion of *a fortiori* arguments again in *FL*. Since complex propositions are the conjunction of two simple propositions, we might expect De Morgan's treatment of the syllogism to begin with simple syllogisms and to move on to complex syllogisms. In fact, he does just the opposite, on the ground that the complex syllogism is easier to understand. In doing so, he refers to the way in which *a fortiori* arguments were handled in *FN*.

> Hitherto the complex syllogism has never made its appearance in a work on logic, except in one particular case, in which it is allowed to be treated as a simple syllogism, though most obviously it is not so. I allude to the common *a fortiori* argument, as in "A is greater than B, B is greater than C, therefore A is greater than C."...
>
> Admitting fully that the quality of the premises,–that which entitles the conclusion to

> be made, as it is said, *a fortiori*–marks this argument out as, if anything, stronger, clearer, and (could such a thing be) truer, than a simple syllogism; yet it is plain that the very additional circumstance on which this additional clearness depends, takes the argument out of a syllogism, as defined by all writers. By beginning with the complex syllogism, and thence descending to the simple one, it will be seen that we begin with cases which present this *a fortiori* and clearer character. (FL, 88–89)

Remembering De Morgan's account of the *a fortiori* argument in *FN*, it appears that he still endorses the attempt to treat such arguments by means of the complex syllogism, though it is just possible that here he is drawing more of an analogy between the two cases. If inferences exhibiting the transitivity of "greater than" can be treated by complex syllogisms, then the doctrine of the abstract copula would seem not to be needed for that case. Many other transitive relations would remain, though, and such inferences for them could not be dealt with as De Morgan dealt with "greater than" in *FN*.

2. THE BICOPULAR SYLLOGISM AND THE COMPOSITION OF RELATIONS

While *FL* contains many hints of things to come, the logic of relations is not central to it. The logical treatment of relations first becomes prominent in "On the Symbols of Logic, the Theory of the Syllogism, and in particular of the Copula," usually referred to as "On the Syllogism, No. II," from 1850. The logic of relations plays a much more central role in this paper than in *FL*; and the account of relations is greatly generalized, to include much more complex operations on them.

The first section of *S2* is titled "On the approximation of logical and algebraical modes of thought," an apt name since De Morgan has stated that the purpose of his paper is to discuss "the application of symbols, and the tendency which such application has to develop what I must call the *algebra* of the laws of thought"(S2, 22). The discussion, which is rather puzzling, has two main points.

De Morgan begins by discussing the formal similarity between different kinds of opposition, claiming that "every pair of opposite relations is undistinguishable from every other pair, in the instruments of operation which are required"(S2, 23). These similarities are not always apparent, because in considering various modes of thought we tend to focus on differences, rather than on similarities. De Morgan then considers eleven pairs of opposites which one encounters in logic, such as affirmative-negative, possible-impossible, and universal-particular. As pairs of opposites they should have similar formal properties. However, De Morgan wants to make the stronger and more puzzling claim that these oppositions are, in a sense, equivalent to each other, in that any one can be selected as basic and all of the others reduced to it. As an example, he tries to show that all logical oppositions can be reduced to the opposition of universal and particular. It is important to see that this reduction is not needed for the formal similarity of the several oppositions, any more than for the formal similarity of, say, gain-loss and ascent-descent. His crucial point, for our purposes, is that the formal similarity of opposite relations suggests a topic for formal development.

> I think it reasonably probable that the advance of symbolic logic will lead to a calculus of opposite relations, for mere inference, as general as that of + and − in algebra. On the advantages or disadvantages of its introduction it would be vain to speculate beforehand. (S2, 26)

A second "approximation" between logic and mathematics consists in certain formal similarities between the process of inference in logic and the process of elimination in algebra.

> Speaking instrumentally, what is called *elimination* in algebra is what is called *inference* in logic. If there be four equations (assertions) in algebra (logic) involving any number of objects of equation (assertion), it is possible from the four to produce one equation (assertion), excluding three objects of equation (assertion) from among those in the originals. The resulting equation (inference) is free from the three eliminated quantities (middle terms). (S2, 27)

Furthermore, there is the similarity that both disciplines rest on breaking down complex sets of eliminations or inferences into simpler ones. In equational elimination, we eliminate a quantity by expressing it in terms of others; and in logic, we infer by describing one object in terms of others. There is a formal similarity between,

(3.14) John met Thomas in the street.
John shook hands with Thomas.
———————————————————
John shook hands with what John met in the street.

and,

$$(3.15)\quad \begin{array}{l} a + x = b \\ y = c + x \\ \hline y = c + b - a. \end{array}$$

This eliminational view of inference pervades De Morgan's treatment of the syllogism, in which the basic rules are all stated in terms of elimination, and in which the entire system of notation is devised to facilitate this.

Even here, however, there is only an approximation between algebra and logic, because there are some formal properties in algebraic elimination that fail for logic (S2, 27). In algebra, equations are perfect identities, so that conversion holds. Thus, if,

$$y = \Phi x \text{ and } z = \Psi x \text{ gives } z = Xy,$$

then

$$z = Xy \text{ and } z = \Psi x \text{ gives } y = \Phi x.$$

In (3.15), for instance, we can recover $a + x = b$ from $y = c + x$ and $y = c + b - a$. In logic the situation is different, since we cannot recover the first premise of (3.14) from its second premise and its conclusion.[3]

De Morgan emphasizes that more complicated forms of logical elimination can be devised as well, such as,

(3.16) John met Thomas.
John and Thomas live on the same street.
John is richer than Thomas.

A person who met one not so rich as him whom he met, and that same person not so rich as him whom he met, live on the same street. (S2, 28)

This inference may appear tortuous, but De Morgan's objective is to extend logic far beyond the narrow bonds of the syllogism, by using eliminational methods. In doing so, he hopes to,

> free logic from being to the higher processes of thought what algebra would have been to its present state if it had discussed no forms more complicated than $x = y$, $y = z$, &c. The world at large is more advanced in constructive process than the books on logic; in which arbitrary separations are declared by old authority to contain all the forms of thought. (S2, 28)

Sections 2–4 of *S2* contain little that is directly relevant to the logic of relations. They present an exposition of De Morgan's notational innovations, including his use of contrary terms and his "spicular" notation, together with a critique of Hamilton's logic and an attempt to render it coherent through what De Morgan calls the "exemplar" interpretation. De Morgan was to use this "spicular" notation, with varying interpretations, in the rest of his logical work. Since we will need to refer to it from time to time, it will be convenient to introduce it now. The basic idea of the notation is simple. Enclosing parentheses, as in "X)" and "(X" show that a term is used universally; excluding parentheses, as in ")X" and "X(" indicate that a term is used particularly. No dots, or an even number of dots, mean that a proposition is affirmative; an odd number of dots, usually one, means that it is negative. The eight propositional forms now become:

X))Y	Every X is Y.
X(.(Y	Some Xs are not Ys.
X((Y	Every Y is X.
X).)Y	Some Ys are not Xs.
X).(Y	No X is Y.
X()Y	Some Xs are Ys.
X(.)Y	Everything is X or Y or both.
X)(Y	Some things are neither Xs nor Ys.

(S2, 35–36)

These are the eight combinations which are notationally possible, of which the last two are non-Aristotelian statement forms. Their presence is one of the greatest novelties of De Morgan's syllogistic. The eight forms are the same as those which he presented in *FL*. When we allow lower case letters to represent contrary terms, a great many other forms may be written, but all of these will be equivalent to one of the eight already given. Thus, "y))x" and "X).(y" are equivalent to "X))Y"; and, "x()y" is equivalent to "X)(Y". We need not consider the simple canon of validity which De Morgan obtains for this system, nor his comments on Hamilton.

De Morgan returns to the doctrine of the abstract copula, and its generalization, in Section 5, "On the theory of the copula and its connexion with the doctrine of figure." This section is of paramount importance for the development of De Morgan's logic of relations. In it, he takes his clue again from algebra, utilizing a conception of algebra which he had long advocated. While we shall discuss this algebraic background of his logic in some detail later on, some indication of it is important at this stage.

> In algebra, as it now stands, the *forms* born and educated in arithmetic have left their parent and set up for themselves. Any meanings which obey certain specified laws may be adopted as the means of giving expression to the forms: and the results must be accepted as true in every instance in which the combinations used are consistent and intelligible under the meanings given. . . .
>
> In my work on *Formal Logic* . . . I followed the hint given by algebra, and separated the essential from the accidental characteristics of the copula, thereby showing the conditions of invention for a copula different from the ordinary one, or for a copula which, being substituted for the ordinary one, shall leave all the forms and conditions of inference unaltered. (S2, 50)

He then proceeds to deal with the comments of an unnamed "learned critic" who has complained that "some of my modes of making the copula are less abstract, none more so, than *is* and *is not*."[4] To rebut this charge, he clarifies his view on the abstract copula.

By an "abstract copula," De Morgan means "a formal mode of joining two terms which carries no meaning, and obeys no law except such as is barely necessary to make the forms of inference follow. Any concrete or actual copula, fulfilling other conditions, is, to the extent of those conditions, less abstract" (S2, 51). In fact, we might well have devised an artificial symbol to stand for the abstract copula, just as we use letters to stand for the terms of a simple proposition. If we use the symbols "–" and "– –" to stand for the affirmative and negative copulas, respectively, the three conditions on the copula which De Morgan presented in *FL* can be stated as:

1. Transitiveness: X–Y–Z = X–Z.
2. Convertibility: X–Y = Y–X.
3. Contrariety: Exactly one of X–Y and X– –Y must hold.

When contraries are allowed, a fourth condition is required,

4. Contraries: Either X–Y or X–y holds for any X.

It should be noted that De Morgan uses " = " as a sign of implication here, as he did in *FL*, even though he will drop it later because of the non-convertibility of implication. He lists fourteen transitive copulas, including "is," "is equal to," and "joins," and he includes five convertible copulas among the transitive ones. Many of the transitive copulas are not convertible, and some convertible copulas, such as "converses with," are not transitive.

Given these basic laws, De Morgan constructs the system of logical laws which follow solely from the formal conditions stated above, and then relates these to the theory of the syllogistic figures. One of his main objectives is to determine just which formal conditions are required for which figures. He begins with systems in which contraries are not present, and he first considers the Transitivity Condition. He calls this condition a "condition of permanence" for the copula, in the sense that any generalization of the copula must contain this principle. This may seem peculiar at first, since at least one common use of "is," the predicational sense, does not satisfy this condition; but, our discussion has shown that this is a natural assumption for him to make.

As he did in *FL*, De Morgan uses the formal properties of the copula in singular propositions to generate the formal laws of quantified propositions (S2, 54). In the simplest singular case,

(3.17) Y gives Z.
X gives Y.
―――――――
X gives Z.

merely expresses the transitivity of "gives" (in the sense of "produces"). If we now let term-letters stand for classes, this yields "Barbara" for "gives":

(3.18) All Y give some Z.
All X give some Y.
―――――――――――
All X give some Z.

De Morgan does not show how the quantified syllogism is obtained from the singular syllogism, but it seems to require instantiation and generalization: consider any X, call it "a"; then since *a* is X, *a* gives some Y, which we may call "b"; then, from the first premise, *b* gives some Z, call it "c." Since *a* gives *b* and *b* gives *c*, and "gives" is transitive, *a* gives *c*. But *a* is any X, so that all X gives some Z.

The use of "opponent forms" of these singular syllogisms gives the following traditional syllogistic forms:

First figure	*Second figure*	*Third figure*
AAA	AEE	EAO
AII	AOO	EIO
		OAO

While it would be needlessly tedious to relate each of these quantified forms to their singular forms, an example will show how transitivity alone, along with instantiation and generalization, suffice. To get 3 OAO (Bokardo), we first note that from (3.17) we obtain the following inference, for singular X and Y:

(3.19) X does not give Z.
X gives Y.
———
Y does not give Z.

From this, we may obtain Bokardo,

(3.20) Some X do not give any Z.
All X give some Y.
———
Some Y do not give any Z.

To show this, let *a* be one of the X's which does not give any Z, and let *c* be any Z. Then *a* does not give *c*. But *a* does give a Y, call it "b," so that *a* gives *b*. By (3.19), *b* does not give *c*. Thus, some Y does not give *c*. But since *c* is any Z, some Y does not give any Z. While (3.20) could be obtained directly from (3.18), when we see that "Some X do not give any Z" is the contradictory of "All X give some Z," it appears that De Morgan intended it to be derived from (3.19).

Having seen that the transitivity of "gives" suffices for seven traditional forms, we should note that it will not suffice for the other forms.[5] Consider the case of 1 EAE (Celarent):

(3.21) No Y gives some Z.
All X gives some Y.
———
No X gives some Z.

Let us now attempt to construct a proof from the singular case as we did in the first two examples. Let *a* be an X, which gives *b*, a Y. Then for any *c* which is a Z, *b* does not give *c*. To obtain our desired conclusion, we need the singular inference:

(3.22) b does not give c.
a gives b.
---
a does not give c.

However, this inference is equivalent to,

(3.23) a gives c.
a gives b.
---
b gives c.,

which is not an instance of transitivity. If "gives" were convertible as well as transitive, though, the second premise of (3.23) could be converted and transitivity then applied. While this cannot be done for "gives," it can be done for "is identical with," from which the other laws of the syllogism which do not involve the use of contraries may be derived.

De Morgan points out that additional syllogisms are allowed if we stipulate that the "correlative copula"–i.e., the converse of the relational copula–may be used as well as the original copula. Thus, the following is a third figure form which De Morgan calls "Darapgi":

(3.24) All the piles supported arches.
All the piles were supported on gravel.

Gravel supported the arches.

The transitivity of support obviously underlies this inference, but the manner in which it does so raises some very important issues. De Morgan claims that this syllogism cannot be reduced to an Aristotelian syllogism. Since this thesis is so important for a full understanding of De Morgan's viewpoint, it will be useful to have his complete statement of it.

> The proposition 'support of support is support' is necessary to the inference, which inference can only be obtained from a sorites, and not then except by help of the *dictum* already quoted. It may be said that this is *more* than an Aristotelian syllogism; I maintain it to be *less*, *if either*. The outstanding copular relation (always implied) of an Aristotelian syllogism is '*is that which is* gives *is*:' of the preceding case, 'support of support is support.' The former demands transitive and convertible meaning for *is*, or

that *is* shall be its own correlative: the latter demands transitive meaning for *support*, and the allowance of its correlative. (S2, 54–55)

To see what point is at issue, let us consider the singular syllogism which is behind (3.24):

(3.25) b supports c.
a supports b.
―――――――――
a supports c.

In saying that this is a sorites, De Morgan appears to be claiming that this inference really has the following form:

(3.26) b supports c.
a supports b.
―――――――――
a is a support of a support of c.
Support of support is support.
―――――――――
a supports c.

The first inference involves the composition of a relation with itself, even though De Morgan's treatment of relational composition is yet to come. The second is, it appears, to be justified by appeal to the *dictum de majore et minore*, which we will discuss at some length in section three. The exact way in which this dictum is to be applied here is somewhat obscure, since the *dictum* is stated as a principle which governs terms, while "support" is the copular relation. Furthermore, the exact formulation of the rule is not clearly applicable here. Nevertheless, the intuitive idea seems fairly clear.

A more basic issue arises at this point: how should the inferential rules be stated? It seems as if De Morgan represented the rules behind such inferences as (3.25) and (3.23) in two quite different ways. In *FL*, he wrote as if there are certain basic forms of inference which simply apply to some copulas and not to others; transitive inferences work just as well for "gives" as they do for "is." He then appeared to tighten up this conception somewhat, by requiring that each such inference should be justified by appeal to a separate rule of inference.

In (3.26), though, we have a quite different strategy. In this case, the transitivity rule has been transmuted into a premise, and the only rules involved are general logical rules–one for the composition of relations, the other for the *dictum*.

Is there really any difference between these two modes of presentation? If so, which is preferable? While axioms can always be converted into rules, and vice versa, it is arguable that logical propriety demands that all the logical machinery be packed into the rules of inference, and that all non-logical principles be put into the premises. From this standpoint, the second mode of presentation would seem to be the best. We shall soon see, though, that De Morgan is not content with the traditional way of realizing this principle, for he will demand that the traditional Aristotelian syllogism be treated in just the same way.[6] For De Morgan, traditional Aristotelian inferences are based on the transitivity of identity, so we would apparently need:

(3.27) B is C.
A is B.
___
A is that which is C.
"Is that which is" gives "is."
___
A is C.

This means that the fundamental rules of inference for the traditional syllogism are relational composition and the *dictum*, combined with a premise stating the transitivity of "is." De Morgan will return to this subject at the very end of *S2*.

While we have been given a hint of the composition of relations, we have been concerned largely with the doctrine of the abstract copula. At this point, though, De Morgan introduces a dramatic addition which we shall call the *doctrine of the bicopular syllogism and the composition of relations* (S2, 55ff). The ordinary syllogism limits itself to one reading of "is" in any given argument; and even the extensions considered so far assume that the copula, whatever it is, is the same in each premise and the conclusion. Furthermore, we have taken the candidate copulas to be at least transitive, if not also convertible. However, there is no reason for these limitations. We may, in fact, use any copulas we please, transitive or non-transitive, convertible or non-convertible, and we may even have different relations in the two premises. The form of such an inference would be,

(3.28) Every Y has a relation to some Z.
Every X has a relation to some Y.
___
Every X has a compound relation to some Z.

De Morgan symbolizes this by allowing "—" to stand for one relation, "......" to stand for another, and "$\overline{......}$" to stand for their compound. The singular case would be:

(3.29) Y——Z
X......*Y*
____
X$\overline{......}$Z,

and examples in the second and third figures can be obtained just as they were with the abstract copula. The result is what De Morgan calls the bicopular syllogism,

> in which the intransitiveness of the individual copulae is supplied by the invention of a compound copula for the conclusion. This is the step by which we ascend to the general theory of the copula: but those who proceed from a general copula to a particular case, will merely see how to read the general conditions in the particular case. (S2, 56)

De Morgan's example of a bicopular syllogism is,

(3.30) John can persuade Thomas.
Thomas can command William.
____
John can control William.,

where by "control" we mean "the process of gaining an end by persuading one who can command."

The crucial issue for the philosophy of logic, of course, is whether the composition of relations is an essentially new operation whose logic cannot be handled by the traditional syllogism. De Morgan immediately considers what he regards as the only alternative. In a somewhat expanded form, this is:

(3.31)

1. John is [one who can persuade Thomas].	*Premise*
2. [One who can persuade Thomas] is [one who can control all whom Thomas commands].	*From Postulate*
3. John is [one who can control all whom Thomas commands].	*From 1,2*
4. Thomas can command William.	*Premise*
5. [One who can control all whom Thomas commands] is [one who can control William].	*From 4, by dictum*
6. John is [one who can control William].	*From 3,5*

Steps 3 and 6 just represent the transitivity of "is," and in a full expansion of the argument they would be expanded into inferences such as (3.27). The use of De Morgan's *dictum* in deriving 5 from 4 is not completely clear, but De Morgan may have had the following in mind:

(3.32) 4a. [One who can control all whom Thomas commands] is [one who can control all whom Thomas commands].
4'. William is [one whom Thomas commands].

---

5. [One who can control all whom Thomas commands] is [one who can control William].

The rule of inference which is used here is the *dictum*, though the complexity of the case illustrates the point, to be made later, that the application of the *dictum* to complex cases is more a matter of intuition than of readily detectible form.

This reduction remains problematic, for De Morgan does not show how the general conception of "control" is to be applied in this specific case to yield step 2. He is alert, though, to a more serious point that is made by this reduction. While this may be a reduction of the bicopular syllogism to the unicopular syllogism, it does not show that it is possible to eliminate the composition of relations entirely. The transitivity of "is," and thus the application of the composition of relations to the identity relation, is essential to this inference in steps 3 and 6.

All this assumes, of course, that inferences like steps 3 and 6 are best thought of in terms of the composition of relations, rather than in more traditional terms. Dialectically speaking, one must point out that De Morgan may think that this construal is natural because it makes this argument form a special case of the more general principle of the bicopular syllogism. But to the extent that the bicopular can be reduced to the unicopular, this type of appeal is diminished. In fact, the defender of the traditional approach might rejoice in this reduction on the ground that it showed that the relational approach was unnecessary, and then construe steps 3 and 6 in a non-relational way.

It is clear at this point that the mathematician's urge to generalize is an important factor in De Morgan's approach. Immediately after attempting this reduction, he says:

The algebraical equation $y = \Phi x$ has the copula $=$, relatively to y and Φx: but relatively to y and x the copula is $= \Phi$. This is precisely the distinction of 'John can persuade Thomas' and 'John is [one who can persuade Thomas.]' The deduction of $y = \Phi\Psi z$ from y = Φx, x = Ψz is the formation of the composite copula = $\Phi\Psi$. And thus may be seen the analogy by which the instrumental part of inference may be described as *the elimination of a term by composition of relations*. For though in ordinary inference the concluding copula is usually identical with those premised, yet it is no less true that the composition must have taken place: X is Y, Y is Z, therefore X is that which (= is) Z. (S2, 56)

The urge to generalize appears again when De Morgan notes that there is much to be learned about the unicopular case by considering it as a special case of the bicopular syllogism,

just as a beginner in algebra sees properties of a × b which would be much clouded by taking his initial examples from a × a. Indeed, the forms of logic, stinted down to the Aristotelian syllogism, much resemble those of an algebra in which all the letters are equal, and all equal to unity. (S2, 57)

Even if the bicopular syllogism can be reduced to the unicopular syllogism, the unicopular syllogism can be illuminated by considering it as a special case of the bicopular. One illustration of this fact, according to De Morgan, concerns the several figures of the bicopular syllogism. Let "——" and ". . . . . ." be two copular symbols, and let "– –" and ". . . . . ." be their negatives. Then through the use of the opponent syllogism, with which we are already familiar, we may obtain the second and third figures from the first figure.

First figure	*Second figure*	*Third figure*
Y——Z	Z——Y	Y$\overline{\ldots}\ \overline{\ldots}$Z
X. . . . . .Y	X$\overline{\ldots}\ \overline{\ldots}$Y	Y. X
X$\overline{\ldots\ldots}$Z	X. . . . . .Z	X— —Z

If the correlative copula is introduced, as with the unicopular syllogism, the full theory of the syllogism may be obtained. In this case, too, the fourth figure is the only figure to require the correlative copula. This may be the source of the widely held view that there is something unnatural about the fourth figure, though, as De Morgan notes, the logician should not be concerned with naturalness, but only with validity (S2, 57).

While the affirmative syllogism exhibits the *composition* of relations, the negative syllogism shows the *resolution* of relations. Thus, instead of (3.31), we might have,

(3.33) Z can command Y.
X cannot control Y.
———————————
X cannot persuade Z.

To show this, we must resolve control into persuasion to command.

De Morgan notes that this resolution is more plausible in the bicopular case than in the unicopular case (S2, 57–58). We do not seem to need the notion of resolution to validate,

(3.34) Z is Y.
X is not Y.
———
X is not Z.

It is most easily construed as the opponent syllogism derived from,

(3.35) Z is Y.
X is Z.
———
X is Y.,

and not as an opponent syllogism obtained from any use of the composition of relations. Perhaps it *is* simpler to treat the transitivity of "is" as a simple rule of inference rather than as an additional premise involving the composition of relations.

De Morgan grants that when "is" stands for "pure identity," the resolution is so "purely formal" that it is reminiscent of resolving 1 into 1×1. However, when "is" only "denotes some agreement, or something transitive and convertible which is not pure identity, I maintain the actual resolution to be part of the process of inference" (S2, 58). If, for instance, "is" means "agreement in color," "Z is Y" and "Z is that which is Y" are different propositions. In this case, the second premise of the inference,

(3.36) Z is Y.
X is not Y.
———
X is not Z.

means that X does not have the color which Y has, for which, as an identity, we substitute "the color which Z has," to obtain the conclusion. In this, the second premise is really, "X is not that which Y is."

De Morgan then explains again the doctrine of the abstract copula.

> Logical writers do not collect their copulae: and their processes seem reducible to common form in no way but this. They postulate that the copulae they use (which they take good care shall be transitive and convertible) have the inferential force of the identity IS: they examine and dwell upon syllogisms of identity, thereby giving the opponents of logic some reason for their scoffs at the syllogism, and return with their results to other copulae, by aid of the postulate. (S2, 58)

The logician's fixation upon identity leads to artificially formulated arguments, in which every type of "is" must be reduced to an "is" of identity. This, in turn, requires that all predicates be of the same logical type as their subjects. Thus, "the rose is red" must become "the rose is a red thing" or "some red thing is a rose." For De Morgan, though, "The admission of relation in general, and of the composition of relations, tends to throw light upon the difference between the invented syllogism of the logicians and the natural syllogism of the external world" (S2, 59). To illustrate this phenomenon, De Morgan considers the following argument:

(3.37) The rose is red.
Red is pretty.
―――――――――
The rose is pretty.

The "logician's pupil" will give this an objective reading:

(3.38) The rose is a red thing.
A red thing is a pretty thing.
―――――――――
The rose is a pretty thing.

The parity between subject and predicate is preserved here by adding "thing" to both predicates. The "logician" does just the opposite:

(3.39) The color of the rose is red color.
Red color is pretty color.
―――――――――
The color of the rose is pretty color.

In this form, the logician "reduces it to the equation of two terms by the comparison with a third." The real form of thought, though, is the composition of relations: even though "is" is used in both premises, it has different meanings in each case. The conclusion, then, expresses the composition of these two senses of "is." We have,

(3.40) The rose is (in its relation to one sense) red.
Red is (in the relation of the perceptions to the judgment) pretty.
―――――――――
The rose is (to one sense that which is to the judgment) pretty.

The distinction of copulae, says De Morgan, is evident from the fact that the first premise is a statement of fact, while the second premise is a matter of taste, so that the "is" in the conclusion is composite.

The line of argument here is obscure, to say the least. De Morgan complains that the first version (3.38) can't be correct, because the second premise is too strong, since it would also mean that a red cabbage is pretty. This raises the question, though, of just what the original argument (3.37) amounts to. It appears to be invalid, since from the fact that red is a pretty color it does not follow that being red makes something pretty. One might even bring out a red cabbage to illustrate the point. This suggests that the original argument is not concluding that the rose, as a whole, is pretty, but only that one aspect of it is pretty–namely, its color. This means that the second version (3.39) should be the preferred version.

But De Morgan's discussion of the second version is also puzzling, for he takes "is" to be used in the same sense in all three occurrences. This sense is, presumably, that of identity. This is an odd reading. A more natural one would take "is" in the first premise to be the "is" of identity; and in the second premise, the "is" of predication. The conclusion could then be construed as an example of the composition of relations, so that it would read:

(3.39a) The color of the rose is red color.
Red color is pretty color.
———————————————
The color of the rose is that which is pretty color.
———————————————
The color of the rose is pretty color.,

where it is emphasized that the "is" in the first conclusion has different meanings in its two occurrences. De Morgan, though, must think that (3.39) is trying to express,

(3.39b) The color of the rose is identical to red color.
Red color is identical to a pretty color.
———————————————
The color of the rose is identical to a pretty color.

In construing the "logician's" (3.39) in terms of (3.39b), De Morgan is holding to his theme that for the logician, all meanings of "is" must ultimately be reducible to identity. De Morgan complains, though, that this does not represent the "real form of thought," for the "is" has two different meanings. It is not at all clear what these two meanings are;

but even granting that the "is" has two meanings, we may well wonder why De Morgan does not select (3.39a) instead. I suspect that the apparent artificiality of translating this argument into "color" terms may have played a role in his preference.

In any case, De Morgan's own version (3.40) is hardly crystal clear. The parenthesized comments, which are De Morgan's own, are so obscure that they do little to remove the problem of the red cabbage. One can only assume that the conclusion, odd as it is, is De Morgan's way of saying that the rose, to the extent that it is red, is pretty. The most interesting point is De Morgan's argument for the equivocity of "is" in this argument, for it is based on the fact that the first premise is a statement of fact and the second a statement of taste. Why should this difference mean that "is" is being used in two different senses? Consider the following argument:

(3.41) All Athenians are Greeks.
All Greeks are handsome.
____
All Athenians are handsome.

The first premise is a statement of fact; the second, of taste. But (3.41) seems to be a clear instance of the transitivity of class inclusion, without any ambiguity of copula at all. The difference between matters of fact and of taste lies in the predicates and not in the copulas. The problem with (3.37) is not in the distinction between fact and taste, but in its conclusion. One cannot argue that the rose is pretty just because it is red; one can only say that one aspect of it is pretty–its color. For this purpose, some version of (3.39) would seem to be preferable. Any exact formulation of such an argument would get us into familiar metaphysical perplexities about the distinction between properties, their instances, and the objects which have them. Such a formulation might well use "is" to mean identity in one premise and predication in the other. This would substantiate De Morgan's claim of the several meanings of "is," without utilizing the very obscure senses required for (3.40). The inference needed for (3.39) could be construed as a simple case of substituting identicals; but it could also be construed, as we did in (3.39a), as an instance of the composition of relations.

In comparing the multiplicity of copulas with the "is" of identity, De Morgan makes a comment which, if pursued, would have resulted in something approaching the full logic of relations (S2, 60). When the "is" of identity is the copula, it will relate, say, just one man with one animal: one man is identical with one animal. With other copulas, though, this restriction need not hold. One book, say, can agree in color with one, or several, or all members of the class of maple leaves. We can thus distinguish between,

One X is L to some Y.

and

One X is L to every Y.

More simply, we can distinguish between "John is L to some Y" and "John is L to every Y." Furthermore, we can make the propositions more complex by quantifying the subject as well to obtain such forms as,

Each X is L to all the Ys.
Each X is L to one or more Ys.
Some X are L to all the Ys.
No X is L to one or more of the Ys.

This extended system, in which both subject and predicate are quantified, allows De Morgan to construct a new interpretation of Hamilton's notation. More importantly, it would have allowed him to construct a much more extensive system in his very important memoir of 1860, in which he confined his attention to the form "X is an L of some Y." We must wait for Peirce to obtain a system which has the form "X is an L of every Y."[7]

So far, our discussion has dealt with a system which does not include contraries. De Morgan concludes his discussion of copulas by considering a system in which both correlative (converse) copulas and contraries are allowed (S2, 63). Here the extended copula will be simply "has relation to," and its correlative will be "has correlation to." De Morgan notes that all transitive relations have transitive correlatives. As examples, he uses "gives to" and "receives from." The basic condition on contraries is the following:

Either X–Y or X–y (but not both) should hold for any X.

This means that a logic developed by analogy with the formal conditions which apply to the copula "is" must meet very specific conditions which many relations do not satisfy. This is not to be lamented, though, for transitivity and convertibility are also important restrictions. For reasons of existential import, we will assume that each X and Y give to something or other, and that each X and Y receive from something. Finally, by analogy with "is," we will assume that no X is such that X–x. From this we can argue as follows: suppose that some X gives to some Y; then this Y cannot give to an x since, by transitivity, this would mean that the X gives to an x. But since this Y must give to something, it gives to an X. Thus, for any X, there are only two possibilities: either that X gives to one or more Ys, each of which then gives to one or more Xs; or, the X gives to one or more ys, each of which gives to one or more Xs. He represents these alternatives by the notation,

X,Y' and Y,X'

and

X,y' and y,X'

respectively. This is not a very perspicuous notation, for it does not explicitly represent the relation, and it suggests that each alternative is the conjunction of two propositions of the same form. If we let "G" stand for the relation involved, use "v" and "w" as variables, and let "a" stand for some X, the first alternative amounts to,

$$(\exists v)(Yv \;\&\; Gav) \;\&\; (v)[(Yv \;\&\; Gav) \supset (\exists w)(Xw \;\&\; Gvw)].$$

The superscript and subscript accents may be combined with De Morgan's spicular notation to yield such propositions as "X,))'Y," which stands for "every X gives to one or more Ys." Unfortunately, De Morgan never fully explains this notation, even though he does state some logical laws using it.

We have now examined the chief ideas in De Morgan's dramatic rethinking of the syllogism: the doctrine of the abstract copula, to be found in *FL*, and the theory of the bicopular syllogism and the composition of relations in *S2*. The latter is a natural generalization of the former, and it provides the motivation for the logic of relations which De Morgan will develop in *S4*. At the same time, De Morgan's new ideas present serious philosophical problems. Before discussing these issues in Chapter Four, we will conclude this chapter with De Morgan's analysis of an important type of relational inference.

3. OBLIQUE INFERENCES AND DE MORGAN'S DICTUM

In *FL* and *S2*, De Morgan discusses a relational inference which is quite distinct from those we have encountered in the composition of relations. Rather than using the composition of a relation with a relation, this other type of inference involves the composition of a relation with a class. Such inferences had been discussed by earlier logicians, such as Aristotle and Jungius, and had sometimes been called oblique inferences.[8] To De Morgan belongs the credit for bringing these inferences to the attention of modern logicians. His discussion of them is both interesting and highly problematic.

De Morgan's example of an oblique inference is familiar to most students of modern logic. In *FL*, he formulates it as,

(3.42) Man is animal.
____
The head of a man is the head of an animal.,

declaring roundly that it "is inference, but not syllogism." It is, rather, a "process which is often necessary, in the formation of the premises of a syllogism, involving a transformation which is neither done by syllogism, nor immediately reducible to it" (FL, 131–32). In *S2*, he reminds us of this example by saying, in a rather combative spirit:

> I gave a challenge in my work on formal logic to deduce syllogistically from 'Every man is an animal' that 'every head of a man is the head of an animal.' From the total absence of attempt to answer this challenge, I conclude that no one has succeeded in whose way it has fallen. (S2, 29)[9]

This quotation helps to put (3.42) in the more perspicuous form,

(3.43) Every man is an animal.
____
Every head of a man is the head of an animal.

Both the premise and the conclusion are now explicitly quantified. It is likely, also, that the uniqueness of the head of an animal, with the apparent use of a definite description in the conclusion, is an accidental feature of this example, and that it would apply equally well to the arms and feet of animals.[10]

If this inference is not syllogistic, what is it? De Morgan's initial characterization of it in *FL* is "the substitution, in a compound phrase, of the name of the genus for that of the species, when the use of the name is particular." Talk of substitution suggests that the form of the argument is,

(3.44) Every head of a man is the head of a man.
Every man is an animal.

Every head of a man is the head of an animal.

In this argument, "animal" is substituted for the second occurrence of "man" in the first premise.

De Morgan continues to use the language of substitution in describing this inference. In both *FL* and *S2* he says that this inference is an instance of what he names the *dictum de majore et minore*, which he construes as a generalization of the traditional *dictum de omni et nullo*. We shall eventually see, though, that this way of putting the matter gives a specious air of conservatism to De Morgan's method of handling this inference. To see this, let us begin by formulating each of these principles.

As we saw in Chapter One, Whately's *Elements of Logic* contains a simple formulation of the traditional *dictum*:

> Whatever is predicated (i.e. affirmed or denied) universally, of any class of things, may be predicated, in like manner (*viz.* affirmed or denied) of anything comprehended in that class.[11]

De Morgan states the same *dictum* in a way which is somewhat odd in its use of "all" and "some," but whose meaning is fairly clear:

> What is distributively affirmed or denied of all, is distributively affirmed or denied of every some which that all contains. (FL, 146)

De Morgan's new *dictum*, though, is quite different. In *FL*, it reads:

> For every term used universally *less* may be substituted, and for every term used particularly, *more*. The species may take the place of the genus, when all the genus is spoken of: the genus may take the place of the species when some of the species is mentioned, or the genus, used particularly, may take the place of the species used universally. Not only in syllogisms, but in all the ramifications of the description of a complex term. (FL, 132)

His formulation in *S2* is different enough to be quoted in full.

> A little consideration suggests as a necessary rule of inference, the right to substitute a larger term used particularly for a smaller one, however used, and a smaller, used in either way, for a larger used universally. What we may affirm or deny of *some* or *all* men, we may affirm or deny of some *animals*: what we may affirm or deny of *all* animals, we may affirm or deny of *all* or *some* men. (S2, 28–29)[12]

There are two features of De Morgan's *dictum* which deserve comment. The first is his claim that this simply generalizes the traditional *dictum*; the second, is the importance of the notion of substitution.

a. A generalized dictum

As a strategy for dealing with oblique inferences, De Morgan's *dictum* has the appearance of a conservative strategy. In part this is because he views its application to them as a preliminary transformation prior to the use of the syllogism, so that it is placed in the same class as immediate inferences such as conversion. More importantly, the new *dictum* appears to be a simple generalization of the old one. In De Morgan's terms: the old *dictum* only allowed the inference from the greater to the lesser, while the new *dictum* allows for an inference from the lesser to the greater as well. Thus, the *FL* version of De Morgan's *dictum* can be divided into two parts:

1. For every term used universally *less* may be substituted.
2. For every term used particularly *more* may be substituted.

On De Morgan's view, the first clause is just the traditional *dictum*, to which the second clause has been added.

The natural question at this point concerns the utility of expanding the traditional *dictum* in this way. While (2) is valid, just as (1) is, it appears that (2) is redundant for the traditional syllogism. As an example, consider 1 AAA:

(3.45) All M are P.
All S are M.
———
All S are P.

The validation of this inference can be considered from the standpoint of either clause. From the standpoint of (1), "S" is substituted for "M" in the first premise to yield the conclusion. Using (2) we can obtain the same result by substituting "P" for "M" in the second premise. Thus, while (2) can be used to justify 1 AAA, it need not be used. And, since the traditional *dictum*, plus the rules for immediate inference and indirect reduction suffice for the traditional syllogism, clause (2) is not really needed for traditional logic.

The new *dictum* does provide an elegant way of handling certain syllogstic inferences. Thus, consider:

(3.46) All A are B.
All C are A.
All B are D.
―――――――
All C are D.

This argument form, which would normally be treated as a sorites, can be construed as a simple application of both clauses of the new *dictum*. Since "A" is used universally in the first premise, (1) allows us to substitute "C" for it; and, since "B" is used particularly in the first premise, (2) allows the substitution of "D" for it.

The main concern of this section, though, is not with elegance of formulation but with logical power, and for this we must turn to non-syllogistic inferences. Let us first consider the oblique inference (3.43) with which we started. As formulated in (3.44), it appears that (2) is necessary, since a greater term is being substituted for a lesser term in a particular occurrence. In fact, (2) is not essential to (3.43), for it can also be construed in terms of (1).

(3.47) Every head of an animal is the head of an animal.
Every man is an animal.
―――――――
Every head of a man is the head of an animal.

As was the case with the syllogistic (3.46), both (1) and (2) can be combined in one inference, as in,

(3.48) Every head of an animal that is Greek is the head of an animal that is Greek.
Every man is an animal.
Every Greek is European.
―――――――
Every head of a man that is Greek is the head of an animal that is European.

In this argument, (1) is applied to the second premise; and (2), to the third premise. We could have got by with (1) alone, by taking as our major premise, "Every head of an animal that is European is the head of an animal that is European," while (2) would have sufficed if we had chosen, "Every head of a man that is Greek is the head of a man that is Greek."

This means that the additional power of clause (2) is not needed for De Morgan's oblique inference any more than it is needed for syllogistic inferences. Does this mean that (2) is useless? Not really, for there are other inferences where it seems to be required. De Morgan himself suggests that those who doubt its importance should consider the complex argument,

(3.49) Certain men, upon the report of certain other men to a third set of men, put a fourth set of men at variance with a fifth set of men.
Every man is an animal.

Certain animals, upon the report of certain other animals to a third set of animals, put a fourth set of animals at variance with a fifth set of animals. (FL, 131–32)

This does seem to be a clear multiple use of clause (2); and, since its major premise states a genuine matter of fact, it cannot be transmuted into another innocuous statement as we did in moving from (3.44) to (3.47).

Simpler cases will work just as well. Clause (1) is required for,

(3.50) All people who love some animal are happy.
All cats are animals.

All people who love some cat are happy.

But (2) is needed for,

(3.51) All happy people are people who love some cat.
All cats are animals.

All happy people are people who love some animal.

Thus, it does appear that clause (2) adds to the power of the traditional *dictum*, though not in all the cases in which De Morgan thought it did.

b. Substitution, not predication

So far we have been construing De Morgan's new *dictum* in a rather conservative manner, involving merely the addition of a second clause to the traditional *dictum*. In fact, the principle is much more radical than this would suggest; and the novelty of this new principle is the source of both its power and its problems.

The traditional *dictum* was stated in terms of *predication*, while De Morgan's *dictum* is stated in terms of *substitution*. This difference, which seems minor at first, is of fundamental importance. Let us first consider the application of this difference to the traditional syllogism, and begin by considering 1 AAA from the predicational standpoint which is contained in the traditional *dictum*. When Whately's formulation of the *dictum* is applied to (3.45), we obtain,

(3.52) P is affirmed universally of all M.
S is comprehended in M.

P is affirmed universally of all S.

We have already seen that when (3.45) is viewed from De Morgan's substitutional standpoint using clause (1), we are merely substituting "S" for "M" in the first premise to get the conclusion.

The difference between these two standpoints seems to have little formal impact on the traditional syllogism; but even here, some inferences which can be performed directly by substitution would require an immediate inference if done traditionally. Consider 2 AEE,

(3.53) All P are M.
No S are M.

No S are P.

The traditional analysis requires a double use of conversion, plus switching the order of the premises, to obtain,

(3.54) No M are S.
All P are M.

No P are S.,

which fits the "denial" form of the traditional dictum. But this is not required for the substitutional form: in (3.53), the conclusion arises directly from substituting "P" for "M" in the second premise.

The ability to substitute directly, as in (3.54), is a mere nicety. The real impact of framing the *dictum* in terms of substitution is that it allows us to apply the *dictum* to items other than full terms. That is, the predicational formulation in the traditional *dictum* applies solely to the full terms that occur as subject and predicate terms of standard categorical propositions. The substitutional formulation goes far beyond this, for it allows us to operate on *parts* of terms as well. This is precisely what happens in De Morgan's treatment of his oblique inference. In (3.44), for instance, "animal" is substituted for "man" within the larger term "head of a man." It is this feature, rather than the addition of clause (2), that allows De Morgan's *dictum* to handle this inference.

De Morgan does not adequately stress the importance of taking this substitutional point of view, though he does touch on the matter. In the first section of *S2*, he draws an algebraic analogy. While algebra can usually deal with elimination of terms between equalities,

> elimination between *inequalities* is sometimes required: and then we know that in $x > y$ our right of substitution is, that we may for x write an equal or greater, for y an equal or a less. In $x > y$, y is used after the manner of a universal term in logic, x after the manner of a particular. (S2, 28)

He then proceeds to state his *dictum* in the form in which we have already quoted it. The analogy here, of course, is with the application of De Morgan's *dictum* to whole terms. The analogue of (3.46) is,

$$\text{(3.55)}\quad \begin{array}{l} x > y \\ w > x \\ \underline{y > z} \\ w > z. \end{array}$$

Rather than thinking of this as the concatenation of three inequalities, we construe it instead as the substitution of "w" for "x" and of "z" for "y" in the first premise.

This algebraic analogy provides a natural transition to the main problem with De Morgan's use of the substitutional model. In the algebraic case, no problems arise so long as we have simple inferences such as (3.55). But the full use of substitution in De Morgan's *dictum* is strictly analogous to the substitution of unequal terms in all kinds of inequalities; and this is fraught with peril, for sometimes it works and sometimes it does not. Take the inference,

$$\text{(3.56)}\quad \begin{array}{l} a - (x - b) > 5 \\ \underline{w > x \qquad\qquad} \\ a - (w - b) > 5 \end{array}$$

Is this inference valid? It is easy to perform calculations to determine whether this substitution holds (it doesn't), but the result would not be something that can be stated as a simple rule of substitution. When "x" is the *full* term in an inequality, there is no problem in framing a rule to govern substitutions from further inequalities in which it enters as a full term; but this does not hold when we allow the "x" to occur in complex terms containing other letters.

An analogous problem arises with De Morgan's *dictum*. When it is applied to traditional syllogistic arguments, no problem arises. But when it is applied to parts of complex terms, serious difficulties ensue. There is a certain irony in this, for it is precisely the case of complex terms which provides the real justification for framing the new *dictum* in terms of substitution. The predicational language of the old *dictum* would have sufficed for the traditional syllogism; and it is only the need to operate on sub-parts of complex relational terms that justifies the substitutional formulation. Yet, it is just such cases that create severe problems for the new *dictum*.

The basic problem is that De Morgan frames his *dictum* by speaking of a term being used particularly or universally, yet he provides no criteria to determine when a term is being used in either of these ways. This does not present a problem for the application of his *dictum* to the traditional syllogism, for particularity and universality are well understood in that context. But the rule breaks down for terms which occur as parts of complex terms, especially the complex relational terms found in oblique inferences. De Morgan provides no criteria to specify which occurrences involve particular uses and which involve universal uses.

This problem can be illustrated by the following simple examples, which are of a sort suggested by Peirce.[13]

(3.57) Every person who loves some human is happy.
Every American is human.
―――――――――――――――――――――――
Every person who loves some American is happy.

(3.58) Every person who loves all humans is happy.
Every American is human.
―――――――――――――――――――――――
Every person who loves all Americans is happy.

In both cases we have moved from the greater term ("human") to the lesser term ("American"). The first argument is valid, while the second is not. The use of De Morgan's *dictum* requires that the occurrence of "human" in the first premise of (3.57) have a universal use, despite its being preceded by the word "some." More strikingly, the occurrence of "humans" in the first premise of (3.58) cannot have a universal use, since the argument is invalid; yet, it is preceded by "all" and is part of the subject term of a universal proposition.

The many possible permutations on these examples show that there are no clear linguistic clues of particular and universal occurrence in complex terms. We are not able first to establish the use of a term, and then to apply De Morgan's *dictum* to determine validity. Rather, we must work the other way around. We first use our logical intuitions to determine the validity of the argument, and then invoke the *dictum* to determine whether specific uses are particular or universal. We would be better advised to treat substitutibility as a criterion of particularity and universality, rather than think of De Morgan's *dictum* as a rule which utilizes independently specified criteria of particularity and universality.

The problems here become more severe as the terms become more complex. At some point, our logical intuitions of validity will vanish, as will our ability to determine particularity and universality. Suppose, for instance, that we consider a variant of the major premise of De Morgan's own example in (3.49):

(3.59) Certain men who are not taller than all men who are not friends with some non-men, put some of the members of a set of men at variance with all members of another set of men.

Can any rule be formulated which would tell us which uses of "men" are particular and which are universal in this odd sentence? With a good ear for language, of course, we could make successive substitutions of "animal" for "man" throughout, and then use our logical intuitions to tell whether the inference held. This, in turn, would tell us which occurrences of "men" involved which kind of uses. But it would be tantamount to abandoning the *dictum* as a logical rule which can be used to determine the validity of arguments.

We must conclude that the very novelty of De Morgan's *dictum* is its undoing. Its primary function is to allow substitution in complex terms, so that oblique inferences, and other relational inferences, can be validated. But it is precisely in such contexts that we cannot utilize the traditional criteria of particular and universal use. And in the absence of some other criteria for particular and universal use, the *dictum* cannot be applied.

CHAPTER IV

THE PROBLEM OF FORM AND MATTER

The conception of logic which we have found in *FL* and *S2*, along with the other logical novelties in these works, were bound to be controversial. The most significant review of these writings was written by Henry L. Mansel and published in 1851. It claimed that most of De Morgan's innovations in *FL* fall outside the domain of formal logic. The chief offender was the doctrine of the abstract copula, for it erred in both its analysis of judgment and its theory of reasoning. It recognized any number of copulas, whereas traditional logic had always insisted on only one affirmative copula. Furthermore, transitivity is not a valid form of inference since it only applies to some relations and not to others; "is joined to" is transitive, while "touches" is not.

Despite Mansel's claim to be reviewing both *FL* and *S2*, his treatment of the abstract copula is based solely on *FL*. He does not even mention the composition of relations. In the second section of this chapter, we show how Mansel seems to have missed one of De Morgan's main points in *S2*. De Morgan's claim is not that an argument using the transitivity of "is joined to" is formally valid; it is just the opposite: traditional syllogistic inferences are materially and not formally valid. Any such inference requires the principle that the identity relation (or the species relation) is transitive, just like the same inference for "is joined to." The traditional copula is as much a material symbol as is "is joined to."

This naturally raises the question of whether there is any meaning to the phrase "formal logic." In *S3*, published in 1858, De Morgan provides a sense for the term "form" which is based on its use in mathematics. A theory of form is "the science of the action of the machinery." Transitivity is a form of inference that is common to "is" and "is joined to"; and the difference between these two copulas is a difference in matter. De Morgan then provides an account of successive levels of form and matter in both propositions and inferences, from which all normal propositions and inferences arise by a process of specification which De Morgan interprets as the successive addition of new levels of matter.

In the fourth section we analyse the points at issue between De Morgan and Mansel. We will suggest that the crucial issue lies in the analysis of judgment. Mansel believed that all judgments are of the subject-predicate form, while De Morgan held that all judgments involve bringing two terms together in relation and are thus relational. Neither seemed to allow, as Thomas Reid had, that there are both subject-predicate and relational judgments.

The chapter concludes with a discussion of Mansel's interesting criticism of De Morgan's treatment of oblique inferences. Mansel claimed that De Morgan's "head of a man" argument is not formally valid, since arguments with the same form have a true premise and a conclusion that is not true–e.g., one which infers that the tail of a guinea pig is the tail of an animal, from the fact that a guinea pig is an animal. The conclusion is not true because guinea pigs have no tails. De Morgan responds with an interesting account of formal validity which takes this problem into account.

1. "SUNDRY PERVERSIONS OF THE SYLLOGISTIC FORM"

Formal Logic received four serious reviews.[1] The review by Joseph Sortain criticized De Morgan's quantitative treatment of logic as found, for instance, in his theory of probability. Sir William Hamilton's long review was limited to De Morgan's views on the quantification of the predicate. The unsigned "learned" notice of *FL* in the *Athenaeum* was primarily an attempt to compare and even reconcile De Morgan and Hamilton, though it did contain, as we have seen, a one-sentence comment on the doctrine of the abstract copula. While these three reviews contribute little to the subject of this study, the fourth review is a very important document for our purposes. It was written by the Reverend Henry L. Mansel, Professor of Metaphysics at Oxford and a person who was closely identified with Hamilton. Mansel's review, entitled "Recent Extensions of Formal Logic," compared De Morgan's *FL* and *S2* with two books by followers of Hamilton; needless to say, Mansel thought Hamilton's work was much more promising than De Morgan's.[2]

It should not be supposed that Mansel's criticisms of De Morgan were either trivial or unsupported. In fact, he raised two very fundamental issues about De Morgan's entire project. One criticism was that De Morgan had subordinated logic to mathematics, that he had made logic an application of mathematics (especially algebra) rather than mathematics an application of logic. This is a defect which he claimed that De Morgan shared with Boole and others, in,

> regarding reasoning as a *computation*, giving a partial and perverted view of the process of thought and its expression by means of mathematical analogies and a mathematical notation, inverting the relation of whole and part, subordinating logic to algebra, and substituting the calculus of inference for the inference of calculation. (M, 93)

There is good reason to believe that this impression is overstated, but it is an issue that we will defer until Chapter Seven, where we will consider the general question of the relationship between logic and mathematics in De Morgan's logic.

The second issue is more fundamental, and it is one which must be addressed in some detail. According to Mansel, *Formal Logic* is grossly mistitled, for all those features which are unique to it are not formal logic at all. Mansel puts his point dramatically by saying:

> The title of Mr. De Morgan's book appears to us a complete misnomer. Under the name of *Formal Logic* he presents us with sundry perversions of the syllogistic form, designed to admit purely material reasonings. It does not seem as if the author had ever asked himself the preliminary question, –What constitutes the matter of thought, and what the form? (M, 105)

Mansel dismisses virtually all of De Morgan's innovations on this one ground, that they introduce material elements into what is supposed to be formal logic.

Mansel illustrates this point by selecting five of De Morgan's methods for detailed criticism, showing in each case how material considerations have entered. Three of these lie outside our concerns: the numerically definite syllogism, the theory of probability, and the use of contraries. Two, however, relate directly to De Morgan's theory of relations. One complaint is that the oblique inference from "every man is an animal" to "every head of a man is the head of an animal" is not a formally valid inference, since it requires the existential assumption of the head of a man. The other criticism, and the one to which we shall devote most of our attention, is an attack upon the doctrine of the abstract copula. This attack goes to the heart of De Morgan's method.

Any discussion of the problem of form and matter in logic must first ask what the distinction between them is supposed to be. It is to

Mansel's credit that he sees this need and that he attempts to provide just such an account. He does so in the belief that "we are acquainted with no work on Logic in which the principle is clearly enunciated, though in most, as far as they are consistent with themselves, it is implied" (M, 97). He invites De Morgan to submit his own version of the distinction so that the public may decide between them, a challenge which De Morgan will take up some seven years later.

Like most traditional logicians, Mansel takes logic to be concerned with concepts, judgments and reasoning; and he finds that the formal-material distinction operates on each of these three levels:

> A concept is an attribute, or collection of attributes, (matter,) representing one or more objects, (form).
>
> A judgment is a combination of two concepts, (matter,) related to one or more common objects, (form).
>
> A reasoning (syllogism) is a combination of two judgments, (matter), necessitating a third judgment as their consequence, (form). (M, 98)

This formal-material distinction is within the realm of the *products* of thought. It is also to be found within the thinking *process*, which "is *formal* when the matter *given* is sufficient for the completion of the product, without any other addition than what is communicated in the act of thought itself. It is *material* when the data are insufficient, and the mind has consequently to go out of the thinking act to obtain additional material" (M, 98). The distinction may then be applied to the three stages of thought to yield *formal conceiving*, *formal judging* and *formal reasoning*.

> If, for example, having *given* the attributes A, B, C, I can think those attributes as co-existing in an object, without appealing to experience to discover what objects actually possess them, this is *formal conceiving*. If, having *given* the concepts, P and Q, I can pronounce "P is Q" without a similar appeal, this is *formal judging*. If having *given* the judgments, "W is X," "Y is Z," I can elicit a conclusion from them alone, this is *formal reasoning*. (M, 98).

Formal conceiving is possible when the attributes do not contradict each other; formal judging, whenever one of the given concepts is contained in the other; and formal reasoning, when the given propositions are connected by a middle term, "under such conditions of quantity and quality that the mere act of thought necessarily elicits the conclusion." Both formal judging and formal reasoning depend upon certain laws, so that, for instance, the validity of 1 AAA depends upon the law of identity. Thinking processes, then, may be either formal or material; but in either case, their products have both a matter and a form. The matter is given *to* the thinking act, and the form is given *by* that act.

This description of the formal-material distinction is puzzling in many ways. Its generality and abstractness combine with an emphasis on mental processes to create this effect. It is a far cry from Whately's very specific language-oriented conception of form. We need not pursue this general conception further, though, for its relevance to Mansel's criticisms of De Morgan becomes apparent only when we get more specific and ask what the formal elements are. Is there just one form of judgment or are there several? How are the forms to be specified? Is there one affirmative copula, or are there several?

Mansel divides judgments into categorical, hypothetical and disjunctive judgments. Our concern will be with categorical judgments, to which he gives a traditional analysis. All categorical judgments are of the subject-predicate form, where subject and predicate are linked by either "is" or "is not." Furthermore, when judgments are put into their true logical form, each copula need be given only one interpretation: "is" just means identity, and "is not" means distinctness. Using Hamilton's quantification of the predicate, 1 AAA is then,

(4.1) All A is [some] B.
All C is [some] A.
———
All C is [some] B.

This is presumably the same as,

(4.2) All A = [some] B.
All C = [some] A.
———
All C = [some] B.

With this interpretation, the validity of (4.2) depends upon the law of identity (M, 100). Using this conception of form, Mansel believes that logic is "the science of the laws of formal thinking" (M, 102). For Mansel, it is extremely important to draw the line between logic and other sciences very carefully: "If a single intruder is admitted from the province of material knowledge, no barrier can be devised which shall not with the same facility give access to all" (M, 103). De Morgan, he claims, has allowed that barrier to crumble throughout his work, and the philosopher of logic must rebuild it.

The doctrine of the abstract copula crosses this barrier in two different ways, both on the level of judgment and on the level of reasoning. Let us consider the judgment first. Mansel rejects De Morgan's theory of judgment on the ground that such "copulas" as "gives" and "brings" are spurious: there are only the copulas of identity and distinctness. In "A gives B," "gives" only appears to be the copula which unites the subject and the predicate terms. The true form of this sentence asserts that "the attribute of giving B is found in the same subject with the attributes forming the concept A" (M, 101). It must be admitted that Mansel's reasons for choosing this elaborate paraphrase are unclear and that a more attractive candidate would be, "A is identical with one of the objects which give B."

Mansel not only denies the legitimacy of these "spurious" copulas; he even denies De Morgan's thesis that "is" can have different meanings depending upon whether one is dealing with names, ideas or objects. Identity of object is the *only* meaning.

> When Mr. De Morgan speaks of the various meanings of *is*, as applied to names, ideas, and objects, he forgets that, in all actual thinking, name, idea, and object are combined. We think of an object, under a concept, which is represented by a sign. When I say "man is an animal," I can mean but one thing, the identity of one at least of the objects thought under each concept. . . . The copula always applies to the object of thought, in that application has but one meaning, and without an object there is no thinking at all. (M, 107)

Mansel's second objection applies to De Morgan's theory of reasoning (M, 106–7). The crux of De Morgan's view is that there is no significant logical difference between,

(4.3) B is C.
A is B.
———
A is C.

and,

(4.4) B gives C.
A gives B.
———
A gives C.,

where "gives" means "produces." The formal law which expresses the transitivity of a relation applies equally to both; and the part of the syllogistic system which depends solely upon the transitivity of "is" applies just as well to "gives." In a similiar vein, the convertibility of identity is the basis for several other moods of the syllogism, and these moods fit "is tied to" just as well as they fit "is."

Mansel points out that the following two inferences are of the same form as (4.4), yet are obviously invalid:

(4.5) The pen touches the paper.
The hand touches the pen.
―――――――――――――――
The hand touches the paper.

(4.6) Achilles killed Hector.
Paris killed Achilles.
―――――――――――――――
Paris killed Hector.

The crucial issue, of course, is similarity of form:

> But how do these examples differ *in form* from "A gives B, B gives C; therefore, A gives C?" He [De Morgan] will tell us that the verb "gives" communicates its action, the verbs "touch" and "kill" do not. But is this knowledge formal or material? Is it derived from the general laws of all thinking, or from a special knowledge of the nature of the actions denoted by the several verbs? If thinking about giving is a different *form* of thought from thinking about killing, there is an end of all general laws of reasoning. The nature of the object thought of must, in all cases, determine the inference. But his fundamental principle is erroneous. The copula, so far as it represents a form of thought, is not ambiguous. (M, 107)

The arguments (4.4)–(4.6) all have the same form. None of them are formally valid, though (4.4) is materially valid. Argument (4.3), on the other hand, is of a different form, for the univocal "is" is part of the logical form of the proposition which contains it.

2. THE MATERIAL COPULA

In a short "Addition" to *S2*, De Morgan had discussed an issue bearing on the problem of form and matter, but in a very novel way (S2, 66–68). Oddly enough, Mansel did not discuss these remarks, or any other specific features of this memoir, in his review, even though the review supposedly included this memoir. As we shall see, Mansel's failure to do so may well have meant that he did not see the full import of De Morgan's position.

In this Addition, De Morgan dealt with the claim, by Hamilton and Mansel, that arguments involving the transitivity of equality could be handled syllogistically. This meant transforming,

(4.7) A is equal to B.
B is equal to C.
———
A is equal to C.

into,

(4.8) Whatever are equal to the same are equal to each other.
A and C are equal to the same (B).
———
A and C are equal to each other.

As we have seen, there are serious technical difficulties with this maneuver, since a strict application of syllogistic rules fails to validate it. De Morgan, though, is not concerned with its technical feasibility. He asks, instead, "What's the point of it?" After all, any inference can be expanded by including the principle which governs it as an extra premise. This, of course, will leave us with a new inference with its own principle which could, in turn, be expanded by including its principle as a new premise.

De Morgan points out that this can be done equally well with the common syllogism. For the sake of a clearer parallel, let us consider a case that is somewhat simpler than that which De Morgan uses. Let the categorical proposition "All X are Y" be expressed relationally as "X is a species of Y." Then 1 AAA becomes,

(4.9) Y is a species of Z.
X is a species of Y.
———
X is a species of Z.

The principle which governs this inference is that a species of a species is a species, so that the inference can be expanded to,

(4.10) Every species of a species is a species.
X is a species of a species of Z.
———
X is a species of Z.

According to De Morgan, (4.8) and (4.10) are exactly parallel: both involve the composition of relations, along with the principle that a certain relation compounded with itself is included within itself. Completely explicit forms of each inference would then be:

(4.11) B is equal to C.
A is equal to B .

A is equal to an equal of C.
Every equal of an equal is an equal.

A is equal to C.

(4.12) Y is a species of Z.
X is a species of Y.

X is a species of a species of Z.
Every species of a species is a species.

X is a species of Z.

In each inference, the first step merely embodies the composition of relations, which is legitimate for any relation whatever. The extra premise which is required for the second step, though, applies only to the particular relation involved, since it expresses the fact that the relation is transitive.

Given the formal similarities between (4.11) and (4.12), why should Hamilton and Mansel demand the expansion of (4.7) into (4.8), yet resist the expansion of (4.9) into (4.10)? Hamilton had insisted that (4.7) was *elliptical*, while Mansel had said that this meant that, as it stood, (4.7) is *material*. For both men, logic should deal with the formal principles which govern valid inferences. This means that rules of inference should be confined to formally valid rules. All of the matter should be placed in the premises.

The same point can be made if we consider the expanded versions (4.11) and (4.12). A traditionalist might agree that both inferences *could* be expanded in this way, if so desired; but he would insist that the additional premises in each argument are of very different sorts. The additional premise for (4.11) is a material principle which states the transitivity of equality. The transitivity of the species-relation, on the other hand, is a formal principle. The transitivity of equality *must* be stated as a premise. The transitivity of the species-relation *may* be stated as a premise, since logical rules can always be converted into premises; but it is not required that we do so and, in fact, some logical principles must always be retained as rules and not as premises. Some logical principles continue to function as rules and not as premises in both (4.11) and (4.12).

Let us now ask how De Morgan would respond to this line of attack. In short, De Morgan agrees that all purely formal principles should be considered as rules rather than as premises, so that a rule of inference properly governs the first step in (4.11) and (4.12). Material principles, on the other hand, may be stated as either rules or premises. Since the transitivity of equality and the species-relation are both material principles, they may each be treated in either way.

De Morgan and his critics agree that the transitivity of equality is a material principle. But they disagree about the transitivity of the species relation: Hamilton and Mansel say that this is a formal principle, while De Morgan says that it is a material principle. What is the basis for this disagreement? The nature of the copula. For Hamilton and Mansel, the copula is a formal sign which forms part of the form of the proposition; for De Morgan, it merely expresses another relation. De Morgan states his position very strongly at the conclusion of the Addition to *S2* when he says:

> There is some want of distinctness in the use of the word *material*, as distinguished from *formal*. . . . matter is opposed by writers, not to form, but to what is recognized as form in the school of Aristotle: the assumption of course being that that school exhausts the forms of thought. Historically speaking, the copula has been material to this day: this I must continue to believe until it be pointed out where the formal conditions have been separated from the matter, and made the instruments of inference, independently of the separated matter. (S2, 68)

It must be emphasized that this is a point which applies not only to the species-relation, but to the copula "is" in general. Even if all predicative uses of "is" can be reduced to the identity relation, identity is still just another relation. And even if "is" should stand for a more clearly predicative sense, as in "The rose is red," De Morgan would contend that it still expresses a relation which is just as "material" as any other relation.

The copula has been material to this day. This is the heart of De Morgan's position, and it means that there is some justice in Mansel's criticisms. According to De Morgan, the validity of the traditional syllogism requires both a formal principle, the composition of relations, and a material principle, the transitivity of the copular- or species-relation. As a result, the traditional syllogism is materially, but not formally valid. This means that there is something puzzling about devoting the bulk of a book entitled *Formal Logic* to the study of the syllogism.

At the same time, it seems that Mansel failed to grasp De Morgan's basic point, perhaps because he dealt only with *FL* and not *S2*. Mansel is most concerned to show that the transitivity of "gives" is not formally valid; and he fails to see that the crucial issue is just the opposite of this. It is De Morgan's claim that the transitivity of the traditional copular relation is only materially valid. This, in turn, is a question of whether the copular relation is just another relation. While Mansel states clearly that it is not, he does not see that the crux of his disagreement with De Morgan lies in just this point.

In defense of De Morgan, it must be noted that he had a different conception of form which made it natural to include the traditional syllogism in a book on formal logic; but for this, we must turn to De Morgan's response to Mansel.

3. DE MORGAN'S RESPONSE

De Morgan waited some seven years to respond to Hamilton's and Mansel's criticisms. This delay may have been due to his knowledge of Hamilton's failing health, which led to Hamilton's death in 1856. In 1858, though, De Morgan returned to logic with a long article entitled "On the Syllogism: III," which included systematic replies to both Hamilton and Mansel. In this memoir, he accepts Mansel's invitation to provide his own characterization of the distinction between form and matter; and in doing so, he proposes a sense of "form" which would make "Formal Logic" an appropriate title after all.

De Morgan begins his reply to Mansel by discussing the general distinction between form and matter. This is necessary, he says, because "To every proposal for a new introduction [into Logic] there is but one answer;–You outstep the bounds of logic, you introduce material considerations" (S3, 75). The form of thought (or the law of thought) is explained by De Morgan analogically, by saying that it "is detected when we watch the machine in operation without attending to the matter operated upon." The distinction may be drawn on several different levels, so that form may itself be separable into form and matter, as may matter.

De Morgan finds the origins of the form-matter distinction to lie in mathematics, though only the logician is likely to worry about it: "the great distinction of form and matter is more in the theory of the logician than in his practice, more in the practice of the mathematician than in his theory" (S3, 78). Mathematicians must continually separate the two in the process of abstraction and generalization, though only in some cases does this mean abstraction of the "instrument from the material." There are various levels of form and matter in mathematics: what is formal in arithmetic becomes material in algebra; and lower forms in algebra become material in more abstract forms of algebra; and so on. The mathematician constantly separates form from matter, but often does not realize what he is doing.

Logic, however, is another story. According to De Morgan, logic has always concerned itself with form, but only with Kant has it been defined as the study of the form of thought. Quoting Hamilton's student Baynes, De Morgan puts this by saying, "Logic bears on its modern banner, The form of thought, the whole form, and nothing but the form."[3] However, this is coupled with a reluctance to accept any forms not recognized by Aristotle.

> The proper reply to every accusation of introducing the material where all should be formal, is as follows. You say this thought or process is material: now every material thinking has its form: therefore this thought has its form. Logic is to consider the whole form of thought: *your* logic either contains the form of this thought, or it does not. If it contains the form of this thought, shew it: if it does not, introduce it. (S3, 79)

De Morgan then applies this general reply to Mansel's views on the generalized copula, the numerically definite syllogism, and the introduction of contraries. We will confine our attention to the copula.

De Morgan begins his reply by restating the fundamental position of *FL* and *S2*. It bears quoting in full.

> In my last paper, as in my work on *Formal Logic*, I separated form from matter in the copula of the common syllogism. The copula performs certain functions; it is competent to those functions; it is competent because it has certain properties, which are sufficient to validate its use, and, all cases considered, not more than sufficient. The word '*is*,' which identifies, does not do its work because it identifies, except in so far as identification is a *transitive* and *convertible* notion: 'A is *that which is* B' means 'A is B'; and 'A is B' means 'B is A.' Hence every transitive and convertible relation is as fit to validate the syllogism as the copula '*is*,' and by the same proof in each case. (S3, 79–80)

In a footnote immediately preceding this section, he had put the point by saying:

Is there any *consequence* without *form*? Is not *consequence* an action of the machinery? Is not logic the science of the action of this machinery? Consequence is always an *act* of the mind: on every consequence logic ought to ask, What kind of act? what is the *act*, as distinguished from the *acted on*, and from any inessential concomitants of the action? For these are of the form, as distinguished from the matter. What is the difference of the two syllogisms above [one involving the transitiveness of equality; the other, identity]? In the first case the mind acts through its sense of the transitiveness of '*equals*:' in the second, through its sense of the transitiveness of '*is*.' Transitiveness is the common form: the difference between *equality* and *identity* is the difference of matter. But the logician who hugs identity for its transitiveness, cannot hug transitiveness: let him learn abstraction. (S3, 79)

De Morgan then applies this general standpoint to Mansel's criticisms of his theory of judgment and his theory of reasoning. We will consider each of these issues in turn.

Mansel had insisted that relational judgments be given a copular form. Thus, "A gives B" must become a judgment such as, "A is an object which gives B." De Morgan rejects this imperative completely. While granting that the original and its translation are "coexistent and coextensive," he insists that they differ in their presuppositions. The translated form assumes that there is something which gives B, and answers the question, "What is it?" In the original form, both objects are treated on the same level, and they are thought of as related by the relation expressed by "gives." The logician's translation will "give as much inference" as the original, but it does not represent the form in which the original is thought. (S3, 83)

De Morgan, then, resists the copular translation of relational judgments. Equally importantly, he insists upon a relational formulation of copular judgments. He shows this by giving a series of propositional forms, moving from the more material to the purely formal (S3, 80). Slightly reformulated, this is:

(4.13)	1.	(Positively true)	Every man is animal	
	2.	--------	Every man is Y	[Y has existence]
	3.	--------	Every X is Y	[X has existence]
	4.	--------	Every X—Y	[— is transitive]
	5.	--------	Every X—Y	[— is any relation]
	6.	--------	α of X—Y	[$0 \leqslant \alpha \leqslant 1$]
	7.	(Probability β)	α of X—Y	[$0 \leqslant \beta \leqslant 1$]

The last two steps allow for the numerically definite syllogism and the theory of probability, innovations which Mansel also condemned strongly for introducing material elements into logic. We are concerned only with Steps 3 through 5, for which we need not consider the prefix "Positively true." For the traditional logician, Step 3 is one of the logical forms, since it contains only formal expressions and variables. For De Morgan, on the other hand, "the copula has been material to this day," and all that matters is that "is" has certain formal properties. "Gives" would have been as appropriate in Step 3 as "is." Even specifying that the copula is transitive introduces a material element, so that in moving toward the "purely formal judgment" even that restriction must also be dropped. Step 5 is the result.

Step 6 suggests an important point, which is independent of the mathematical form in which De Morgan states it. The traditional syllogism does not deal with the "purely formal" judgment, but rather with several forms of categorical propositions. The purely formal judgment should incorporate the fact that there are several possible quantifiers as well as positive and negative copulas. This means that the pure form of judgment for traditional logic should be "*QX–C–Y*," where "Q" can be one of the quantifiers, and "C" the affirmative or negative copula. If De Morgan were dealing only with the traditional quantifiers, and if he were to drop references to probability, his "purely formal judgment" would be "*QX–R–Y*," where "R" stands for any relation whatever. For our purposes, this is the crucial difference between De Morgan and the traditionalist on the issue of the logical form of judgments.[4]

This analysis of propositions is extended to the analysis of reasoning. Just as there are levels of form and matter in the analysis of judgment, so there are levels in the analysis of inference. The materiality of inference is once more emphasized, when De Morgan says that, "every inference which is anything more than pure symbolic representation of inference is due to the presence of something material" (S3, 81). The sequence which he appears to have in mind is the following one, where, for simplicity, the probability operator and the quantifier have been deleted. We use "RS" to represent the composition of the relations R and S.

(4.14) 1. X is Z
W is X
―――――
W is Z

2. X–R–Z
W–R–X
———
W–R–Z

3. X–R–Z
W–R–X
———
W–RR–Z

4. X–S–Z
W–R–X
———
W–RS–Z

5. Y–S–Z
W–R–X

From the two "purely formal" judgments in Step 5, nothing can be inferred, other than their conjunction. If "Y" is replaced by "X," we get the valid inference in Step 4. For De Morgan, this is a material step: "it is now seen to be of the matter of our system that the second subject is the first predicate." De Morgan moves directly from Step 4 to Step 1, but it seems consistent with his approach to interpolate Steps 3 and 2. The result is the material inference which represents the transitivity of the copula.

De Morgan has extended the generality of logic from the terms to the relations between the terms. He characterizes this innovation as follows:

> In common logic, the objects of inference, being terms expressed in general symbols, are void of matter; the relations between them, and the modes of inference, are material: I speak of logic as it is. Many relations have a common form: the logician cannot yet see that when many cases, no matter what, proceed upon a common principle, his concern is with that principle. It is his business to apprehend the principle and to shew, as to the *modus operandi* of the mind, how containing cases severally contain it, and apply it. (S3, 81–82)

4. THE ISSUES

Let us now address the question of whether De Morgan has, in his zeal for the abstract copula, introduced material considerations into "formal" logic. To begin with, it should be noted that the Steps 3 and 4 of

(4.14), which represent the composition of relations, are both formally valid. To the extent that De Morgan is dealing with the *composition of relations* and the *bicopular syllogism*, he has expanded the borders of formal logic, rather than admitting intruders from the outside. This is important because of Mansel's concentration on the doctrine of the abstract copula and his neglect of the composition of relations. It is important, also, because De Morgan's *logic* of relations in *S4* (1860) will consist largely of formally valid inferences.

At the same time, Mansel's concern over the *doctrine of the abstract copula* is understandable. Even when embedded within the composition of relations, it must assume that the transitivity of a relation is given, either as a rule or as a premise. It cannot be stated as a rule, since it is not formally valid. Logic would be saddled with a separate rule for each transitive relation, none of which is formally valid. A science containing such rules cannot be called a *formal* logic.

Mansel would insist that the transitivity of a relation should be stated by using a materially true premise. But then we encounter De Morgan's thesis that the traditional copula "has been material to this day"–that is, that if we require a premise for the transitivity of "gives," we should equally require one for "is." Mansel, of course, would claim that De Morgan has failed to grasp the essential difference between "is" and "gives"; the former, but not the latter, is part of the form and not the matter of a proposition.

De Morgan would resist this criticism on two grounds. The first concerns the meaning of "formal," as it reflects the distinction between form and matter. For De Morgan, form is to be found when we pay attention to the "machinery" and ignore the matter being worked upon. In this sense, all transitively relational inferences have the same logical form, even though some are valid and others are not. In doing formal logic, one is pointing out the formal laws which apply to some relations. We might put this point by saying that De Morgan is concerned with the *formal properties* of relations; and transitivity is a formal property which "gives" possesses but "touches" does not.[5] Thus, there is a sense in which De Morgan would insist that he is doing formal logic, even though he claims that many inferences often thought to be formally valid are only materially valid.

The second point of controversy cuts deeper. Traditional formal logic has been centered on the syllogism, a topic to which De Morgan devoted the greater part of his logical energies. Yet, according to De

Morgan, traditional syllogistic inferences are *materially and not formally valid*, for the transitivity which they embody is no different from that of "gives."[6] Mansel would regard this view as preposterous. Arguments based upon the transitivity of the identity- and species-relations are formally valid. Their validity does not rest upon being seen as instances of the formally invalid schema for transitivity.

The dispute can be clarified by noting that De Morgan insists that all inferences be expressed in the following form:

(4.15) X–S–Z
W–R–X
____
W–RS–Z (Rule 1)
RS is included in T
____
W–T–Z (Rule 2)

This is the form that is exemplified, for instance, by De Morgan's own example of the composition of relations–i.e., that in which the ability to persuade one who can command is the ability to control. Two rules of inference are always used. The first rule merely embodies the composition of relations; in fact, it could be taken to explain what the symbolism for relational composition means. The second principle is an analogue of the traditional *dictum*, except that it applies to relational inclusion rather than class inclusion.

When De Morgan comes to view the abstract copula in the light of the composition of relations, transitively relational inferences become instances of this general form.

(4.16) X–R–Z
W–R–X
____
W–RR–Z (Rule 1)
RR is included in R
____
W–R–X (Rule 2)

The priority of form (4.16) requires that the transitivity of R always be stated as a material premise; and this holds whether R is "gives," "is identical to," or "is a species of."

Mansel would agree with De Morgan that the transitivity of "gives" must be stated as a premise; but he would deny that this should be done for "is identical to" or "is." In making this complaint, Mansel would seem to have a point. Traditional syllogistic inferences *can* be twisted into the relational form (4.16); but why *should* they be? Why not apply the traditional *dictum* directly to syllogistic inferences in their usual form, rather than be forced to a relational formulation of the inference with its own analogue to the *dictum*?

We can make this point more vividly by noting that Rule 2 in (4.16) is a primitive rule for dealing with inclusion between *relations*. In the next chapter we will see that De Morgan explicitly uses the transitivity rule for relational inclusion–i.e., that if R is included in S and S is included in T, then R is included in T. Why should the transitivity of class inclusion be forced into a premise when the transitivity of relational inclusion is taken as a basic logical rule? We could not require that the transitivity of relational inclusion be stated as a premise, lest we be off on an infinite regress. Why, then, should we not take the transitivity of class inclusion as a logical rule?

By now it should be apparent that the crux of this dispute lies not in the analysis of reasoning but of judgment. Mansel has adopted the traditional subject-copula-predicate analysis of judgment, with the added stipulation that the copula is always identity or distinctness. De Morgan, on the other hand, is proposing a new relational analysis to replace the traditional account. This means that he both rejects the copular analysis of relational judgments and requires a relational analysis of subject-predicate judgments.

We can make this comparison in a slightly different way. Both Mansel's and De Morgan's accounts would have been clearer if they had seen that talk of form in logic depends upon criteria for determining when two propositions (or arguments) have *the same form*. This notion has traditionally been defined by dividing the words of a language into two kinds, the content words and the logical words. Then two propositions (or arguments) have the same form if it is possible to convert one into the other by replacing the content words of one with the content words of the other, in a one-one manner.[7]

Since Whately's notion of form is consistent with this account, it is puzzling that neither Mansel nor De Morgan was very clear about it. But the difference between them can still be cast in its terms. Consider the two sentences,

(4.17) Every X is a Y.

Every X loves a Y.

Mansel would say that the second of these sentences would have to be transformed into "Every X is something that loves a Y" in order to see that the two propositions have the same logical form. This is a sentence of the form, "Every X is a Z." For De Morgan, on the other hand, the two sentences already have the same logical form, given by "Every X–R–Y." The crux of the dispute is that "is" is a logical word for Mansel and a content word for De Morgan. We might say that for De Morgan, *ising* and *loving* are just two different relations.

What are we to make of De Morgan's relational account of judgment? Has he not just exchanged one Procrustean bed for another? Mansel, following the tradition, had insisted upon putting relational propositions into subject-predicate form. De Morgan has done just the opposite, by construing subject-predicate propositions relationally. It is one thing to extend logic to include relations; it is quite another thing to think of subject-predicate propositions as relational propositions.

Both Mansel and De Morgan are monists in insisting that all categorical propositions have the same general structure. For Mansel, it is copular; for De Morgan, relational. Is not a pluralistic view more natural? Why not allow judgments involving relations with different numbers of argument places to be of distinct types? This is a suggestion which need not wait for Frege and Peirce, since Thomas Reid had already made it in contrasting the relational inferences of mathematics with the qualitative judgments of everyday life.[8]

One way of making this point is to note that in standard symbolic logic, the copula is buried, as it were, so that the temptation to construe it as a relation is avoided. It is not obviously present in "P(a)" or "R(a, b)," though it may be lurking somewhere in the parentheses or the syntactical structure. In either case, though, it is hardly like one of the descriptive constants. This avoids the problems of both Mansel and De Morgan. It appears as if they both treated the copula as a relation: it's just that Mansel thought it to be a formal relation, and De Morgan thought it to be material.

It seems that any notation which represents the copular "relation" as a relation runs the risk of "ballooning" propositional forms into an arbitrary number of terms. If a copular relation is needed to unite "P" and "a" into "P(a)," why isn't a similar relation needed to unite "R," "a," and "b" into "R(a,b)"? If something must unite the elements of a proposition, it must be of a radically different sort than the elements which are united. From this standpoint, we are far better off not to represent the copula as a relation at all, but rather to incorporate it implicitly into our notational structure.

What might be said for De Morgan's position in the face of these objections? Since De Morgan more often states his position than defends it, some speculation is required. However, something can be said. To begin with, we may approach De Morgan within the context of his time. He undoubtedly was influenced by the fact that for many logicians, Mansel included, the copula did express a relation–the identity relation. If the copula can express this one relation, why should it be thought of as part of the formal structure of the proposition, and all other relations consigned to the matter? There is a peculiar inelegance about having just one formal relation and a multitude of material relations. Furthermore, there are close formal similarities between the copula and other relational expressions. This goes beyond mere syntactical similarity to the fact that only the formal properties of the copula are needed for its inferential role; and it shares these properties with a great many other relations. Nothing that is specific to the meaning of the copula need enter the traditional theory of the syllogism.

More broadly, how should the relation between subject and predicate be represented? Our basic concern is that there is no more reason to represent this by a relational symbol than there is to represent the relation between a relation and its terms by a relational symbol. Yet, De Morgan does the former and not the latter. He does so because he would deny the parallel between predicates and relational expressions. To see why this is so, we must take a glimpse at the positive account of judgment which is given in Section II of *S3*. It is here that we find De Morgan's relational account of judgment.

> When two objects, qualities, classes, or attributes, viewed together by the mind, are seen under some connexion, that connexion is called a *relation*. . . .
>
> A *proposition* is the presentation of two names under a relation. A *judgment* is the sentence of the mind upon a proposition, true, false, more or less probable. . . .
>
> The distinction of *subject* and *predicate* is the distinction between the *notion in relation* and the *notion to which it is in relation*.
>
> When X has a relation (A) to that which has a relation (B) to Y, X has to Y a *combined relation*: the *combinants* are A and B. (S3, 119)

This is a general characterization of the proposition and of judgment. A special case of this involves relations which arise from the notion of naming, the onymatic relations. The essential characteristic of the traditional syllogism is that it is limited to onymatic relations. Looking at logic relationally allows us to incorporate the traditional syllogism into this broader framework. Our success in doing so both illuminates traditional logic and provides support for the broader theory.

> The *syllogism* is inference of the relation which exists between two terms, as a necessary consequence of their relations to the same third, or *middle*, term. When the relations are *onymatic*, so may be called the syllogism.
>
> The perception of the validity of a syllogism is the perception of the combination of two relations into one. (S3, 130)

It may require some effort to peek behind our Fregean curtain and think of propositions as De Morgan did. According to Frege, the concepts which both predicates and relational expressions express are alike in being "unsaturated." They are, as it were, abstract entities with holes in them, and they differ only in the number of holes they have. In forming a proposition, these holes are filled by terms, so that nothing else is needed to knit together the proposition.

For De Morgan, on the other hand, the essential fact about the proposition is that it brings two names together under a relation. This is obviously done with straightforwardly relational propositions such as "A produces B." But subject-predicate propositions also contain two terms; and if there is to be a proposition, those terms must somehow be related. And this is just what the copula does.

This rejects the Fregean parallel between subject-predicate and relational propositions. The unity of the proposition arises from two terms being brought together by a relation. Since the subject and the predicate are the terms for a proposition, there must be some relation which binds them together to obtain a proposition. In ordinary relational propositions, this binding is just what the relation does; the function of the relation *just is* to unite terms into a proposition. As such, it does not make sense to ask for an additional relation to relate the relation to its terms.

Thus, it appears that the copula "is" has three quite different functions in most assertions:

1. It stands for a real relation, which may be identity, species, membership or exemplification.
2. It provides for the unity of the proposition.
3. It expresses a judgment as well as being part of the proposition which is judged upon.

While this is true of "is," it is equally true of any other relational expression. De Morgan thought that what matters about the identity relation, as far as the analysis of the proposition is concerned, is the mere fact that it is a relation–not that it is identity. It is the means by which two terms are united into a proposition. This fits nicely with the claim, which we have already encountered, that only the formal properties of the identity relation are relevant to the theory of reasoning, not the identity relation itself.

The most obvious problem with this view of the proposition is that it does not seem general enough. If we can unite two terms into a proposition by relating them, why not three or four or ten terms? De Morgan's concern with the relational syllogism seems to have precluded this generalization; but there is no reason in principle why it could not be made. For this, though, we must wait for Frege and Peirce.

We have now seen that De Morgan's relational analysis of judgments and inferences goes much deeper than Mansel had supposed. It suggests that the fundamental form of logic should be the logic of relations, something which did not exist at the time De Morgan wrote *S3*. De Morgan would take the first steps toward formulating such a logic in *S4*, to be published two years later. This will be the subject of Chapter Five. We will conclude the present chapter by noting an intriguing dialogue between Mansel and De Morgan on De Morgan's treatment of oblique inferences.

5. HEADS AND TAILS

In Section 3.3, we discussed De Morgan's claim that the following inference cannot be validated syllogistically:

(4.18) Every man is an animal.

———————————————

Every head of a man is the head of an animal.

For De Morgan, this is a propositional transformation and not a syllogistic inference, even in his relational sense of the syllogism. It is validated by the *dictum de majore et minore*.

This inference is Mansel's first example of De Morgan's tendency to let material considerations intrude into formal logic. Mansel claims that this argument is not formally valid, because other arguments of the same form do not yield a true conclusion from a true premise. His example is formulated in the style of De Morgan's treatment in *FL* rather than *S2*:

(4.19) A guinea-pig is an animal.
―――――――――――――――――――――――――――
The tail of a guinea-pig is the tail of an animal.

The truth of the conclusion assumes that guinea pigs have tails, which they do not; but even if they did, this would be a fact of natural history and not of logic. The same is true of the original (4.18). Thus, "the consequence is therefore a special inference, *gained from our material knowledge of the thing thought about*, not a general inference *necessitated by the universal laws of thinking*" (M, 106). He then cites a passage from Wolff's German logic to show that this inference "has not been accidentally neglected," on the ground that the premise does not "directly excite" [Wolff's phrase] the conclusion in the mind.

This is an interesting objection, but it is problematic. For one thing, it is not clear whether Mansel has fairly represented De Morgan's argument in using "the" rather than "every." On some accounts, the former carries existential presuppositions that the latter lacks. I am inclined to think, though, that this difference is not a serious one in this context, since De Morgan himself moves from one to the other without comment. Even with "every," the presumption of existential import would require that there be at least one tail of a guinea pig. But this still seems very far removed from the point at issue. The problem of existential import which Mansel has raised is a real one; but it could be met in several ways: by denying that the conclusion has existential import, or rephrasing it in hypothetical form, or adding a premise which asserts the existence of men's heads. While any of these modifications would meet the objection, none of the revised versions of the argument could be validated syllogistically, so De Morgan's point would still stand.

Interestingly enough, De Morgan does not answer this objection in *S3*, and he only refers to it in passing in *S4* (216). In his article on "Logic" in the 1860 *English Cyclopaedia*, he replies by providing a definition of "formal consequence" which takes this problem into account (L, 252–253). He proposes that a consequence is true by its form just in case "it can be refused admission by impossibility in its matter . . . and by impossibility we mean incompatibility with the conditions of the universe understood." Thus, the impossible will vary with the universe of discourse which is understood. If it is "the whole sphere of possible thought," then the impossible is that which contradicts itself; if it is the actually existing things on earth, it is that which does not exist on earth, such as a rational quadruped; if it is the domain of animal life, the impossible would include such things as a stone. Thus "A is A" is true by its form because we can only refuse to assert it when A is impossible within our universe of discourse.

This account is not very clear, but De Morgan's intent is shown better when he deals with formal consequences. If X is an animal, then the tail of X is the tail of an animal, so long as X has a tail. We may say, then, that,

(4.20) Every X is F.
―――――――――――――――
Every R of X is the R of an F.,[9]

is a formally valid argument because, no matter what X, F and R might be, the truth of the premise will necessitate the truth of the conclusion, *as long as* the subject-term of the conclusion denotes something, in whatever universe is understood. In De Morgan's terms, "we know the *consequence* to be necessitated by the laws of thinking, because we must go to impossible matter, we must make the tail of X a non-existence, before we can refuse to assert it" (L, 253). A non-denoting subject-term forces us only to refuse to assert a proposition, not to deny it. "A guinea-pig, for instance, puts this proposition out of the pale of assertion, and equally out of that of denial; the tail of a non-tailed animal is beyond us." Presumably, it is beyond assertion or denial just because it is neither true nor false.

CHAPTER V

THE LOGIC OF RELATIONS

We have now examined the philosophical framework surrounding De Morgan's views on relations, and we have also seen how these views show the need for a logic of relations. In this chapter, we will discuss De Morgan's central contribution to the logic of relations, which he published in 1860 under the title, "On the Syllogism: IV and on the Logic of Relations." In this classic memoir, De Morgan moves beyond his relational analysis of the syllogism and the bicopular syllogism to something that may justifiably be called a *logic* of relations: that is, the specification and systematization of previously unrecognized valid forms of relational inference.

After a section in which we point out the familiar philosophical framework in *S4*, we outline the core of De Morgan's logic of relations. Notations for the contrary and the converse of a relation are introduced, along with that for the composition of relations. To this are added notations for two new ways of compounding relations, the signs of "inherent quantity" to represent "L of every M" and "L only of M." The laws of this logic concern inclusions and equivalences between relations. Some of the main laws of this system are stated, though its precise and complete formulation will be delayed until Chapter Eight. De Morgan's proof of "Theorem K" will be emphasized, for it comes the closest to being a formal proof of all the results in this memoir.

De Morgan develops the general logic of relations only to the point where it can be used for his familiar syllogistic purposes. This means that he is especially interested in relations which are convertible and/or transitive, and we next turn to his discussion of the laws which govern these types of relations. Laws for convertible relations are stated, and De Morgan's puzzling suggestion that every convertible relation is reflexive is analysed. De Morgan's ability to discover new logical principles is especially impressive in his discussion of transitive relations, where he finds such laws as, "A non-ancestor is always an ancestor of none but non-descendents."

The syllogistic motivation becomes even more apparent in section four. We have seen that De Morgan claims that the traditional syllogism deals merely with relations between classes, so it can be considered a special case of the bicopular syllogism and the composition of relations. This suggests a formal theory of the bicopular syllogism, which De Morgan develops, complete with figures and "phases." It is in this analysis that the need for the new forms of inherent quantity becomes apparent.

Section five discusses a potentially significant extension of the logic of relations, though De Morgan only pursues it to the extent needed for his syllogistic theory. Relational terms can be combined not only with singular terms ("lover of John") and relational terms ("lover of a friend of"), but with class terms as well, as in "lover of a man." The result is a form of relation-class composition analogous to the composition of relations. We will see how De Morgan deals with a limited selection of such cases, and we will note his need for a notation to represent "lover of every man" and "lover of none but men" as well as the one which he uses for "lover of some men."

The most specific form of relational syllogism is found in De Morgan's account of the "limited unit syllogism." This is a bicopular syllogism in which, for some transitive relation L, either L or the converse of L occurs in each premise. De Morgan's laws for transitive relations provide the foundation for these inferences.

We conclude this chapter by considering De Morgan's conclusions about the types of reasoning which occur in ordinary life (delaying his discussion of mathematical reasoning until Chapter Seven). He asserts that the traditional syllogism is used infrequently, and that the unit syllogism, involving the combination of relations, is much more frequent. But far more frequent are inferences involving the composition of terms (similar to the Boolean algebra of classes), traditional propositional transformations (e.g., contraposition), and relational transformations (including oblique inferences).

1. PHILOSOPHICAL PRELIMINARIES

The scope of *S4* is limited in some respects, but it represents a very significant extension of logic. The forms of relational inference which it recognizes are neither common nor obvious; and De Morgan testifies to the "confusions and blunders" through which he arrived at this new form of logic.

The difficulty and unfamiliarity of this material make it all the more puzzling that De Morgan's published and unpublished papers show almost no trace of his labors in producing this important memoir. No published work reveals any preliminary version of these ideas, and there are no preliminary drafts of the memoir still in existence. He never mentions it in any of his unpublished papers and tracts, and he does not discuss it in his correspondence, other than to give a few examples of the inferences he has in mind. Furthermore, he almost never refers to the subject in his published and unpublished papers and letters written after the memoir.[1] A thorough study of all his papers conveys the impression that the memoir was written quickly and then forgotten; yet, one cannot believe that De Morgan's intuitive familiarity with tongue-twisting relational inferences was achieved easily, or that the labor in achieving it was quickly forgotten.

The crux of this memoir lies in the forms of inference which it recognizes and systematizes. This is embedded in a philosophical framework which is largely familiar to us by now. De Morgan begins by lamenting the logicians' neglect of the "laws of thought connected with the notion of relation"; and he notes that the logician "places it among those heterogeneous *categories* which turn the porch of his temple into a magazine of raw material mixed with refuse"(S4, 208). Language and thought have interacted, he claims, to produce a language that is more able to express sophisticated concepts than its predecessors; and logicians neglect the full resources of ordinary speech when they appeal to common language to support their restricted views of the nature of judgment and inference. In fact, they greatly underestimate our logical powers.

> I affirm that all the difference between Aristotle or Occam and the lowest of the noble savages who ran wild in the woods is only part, and I believe a very small part, of the development of human power. If the logician could leaven his own mind with a full sense of what his foregoers did for thought and for language, a spontaneous admission would grow that if these same foregoers had worked themselves into the same familiarity with relation in general which they obtained with what I call *onymatic* relations, and still more if they had cultivated those yet wider fields which lie beyond, the common language would have now possessed facilities on the want of which he founds his assertion of the sufficiency of the old logic. (S4, 210)

Such a logic would allow us to reason about the unfamiliar with a power more nearly equal to that with which we reason about the familiar.

De Morgan then attacks the view of logic which limits the forms of propositions to affirmations and denials of identity, and in which all other relations are absorbed by the predicate. This view claims that all inference is reducible to three laws:

Identity	A is A.
Non-contradiction	Nothing is both A and not-A.
Excluded Middle	Everything is either A or not-A.

De Morgan denies that these forms suffice for the traditional syllogism, and claims that two other forms are necessary:

Convertibility	If A is B, then B is A.
Transitivity	If A is B and B is C, then A is C.

While De Morgan grants the soundness of the three traditional laws, he does not agree that only inferences which satisfy them are valid inferences (S4, 214). In particular he claims that all attempts to derive convertibility and transitivity from the three traditional laws implicitly assume that identity is convertible and transitive. He expresses some hesitation in this claim, though, because of the lack of precision with which such derivations are normally presented. He can only report that his own attempts at doing so have been circular and challenge the opposition to produce a non-circular demonstration.

The derivability of the convertibility and transitivity of identity from the three traditional laws is an issue of great importance, for two reasons. First, De Morgan had long held that the traditional syllogism is based on these two principles, so that failure to derive them should mean that the traditional syllogism cannot be derived from the three traditional laws. Second, this opens the door once again to De Morgan's view of logical form. If all the laws of the syllogism follow from convertibility and transitivity, then its laws should hold for any relations which have these two properties. The restriction of the copula to "is" becomes arbitrary.

De Morgan's hunch that convertibility and transitivity cannot be derived from the three traditional laws is certainly correct. Non-contradiction and excluded middle are just quantified versions of laws of propositional logic, so that the only specifically relational law of these three is the law of identity, which merely states that identity is a reflexive relation. The question is thus whether every reflexive relation is convertible (symmetrical) and transitive. This is obviously false, as can be seen by considering the relation of loving in a world of self-lovers.

Limiting logic to the traditional syllogism means limiting ourselves to one class of relations, the onymatic relations. This is as unreasonable as limiting arithmetic to counting–even more so, says De Morgan, since all arithmetic can be reduced to counting, while not all valid inferences can be reduced to the syllogism, as oblique inferences show. Furthermore, the considerations concerning form and matter from *S3* show that there is no reason to demand that inferences involving, say, the transitivity of non-onymatic relations need be reduced to those which use only onymatic relations. The supreme form of the syllogism remains, as it was in *S3* (S4, 218):

> There is the probability α that X is in relation L to Y.
> There is the probability β that Y is in relation M to Z.
> ---
> There is the probability $\alpha\ \beta$ that X is in relation L of M to Z.

2. GENERAL LOGIC OF RELATIONS

The core of De Morgan's memoir is its middle section, in which the new logic of relations is described (S4, 220–27).[2] It is important to realize that its objectives are very limited. While many general laws of relational logic are outlined, they are developed only as fully as the theory of the syllogism requires. Thus, while the notation is general, and while several extensions are hinted at, the theory itself is grounded in the theory of the syllogism from which it arose. De Morgan's notation for his new logic is not very perspicuous, and its principles are not developed in a clear axiomatic way. In Chapter Eight we shall present the system in a clearer manner, but for the moment we shall stick with De Morgan's own formulations.

De Morgan's notation is based on the interesting observation that in the logicians' use of "is," two different functions are being performed: in "A is B," it stands both for the relation of identity and for the affirmative judgment. In "A is not B," however, it stands only for identity, with "not" representing the act of denial. De Morgan's notation, though cumbersome, makes a systematic effort to separate relation and judgment. Thus, "X. .LY" is the affirmative judgment that X stands in the relation L to Y, while "X.LY" is the negative judgment that X does not stand in the relation L to Y. In addition, lower case letters are used to stand for the complements of relations.

The fundamental notation for De Morgan's general theory of relations is the following:

X. .LY	X is one of the Ls of Y.
X.LY	X is not one of the Ls of Y.[3]
X. .L^{-1}Y	Y is one of the Ls of X (Converse).
X. .lY	X is not one of the Ls of Y (Contrary).
X. .LMY	X is one of the Ls of one of the Ms of Y (Composition).
X. .LM′Y	X is an L of every M of Y (Inherent Quantity).
X. .$L_{,}$MY	X is an L of none but Ms of Y (Inherent Quantity).
L))M	Every L of a Z is an M of that Z (Relational Inclusion).
L‖M	L))M and M))L (Relational Equivalence).

Conversion and contrariety are the two one-place operations on relations. The notation for the composition of relations merely expresses what De Morgan had long emphasized in his writings on the syllogism. It is in the two signs of "inherent quantity" that genuine novelty appears; in fact, De Morgan says that "until this suggestion arrived, all my efforts to make a scheme of syllogism were wholly unsuccessful"(S4, 221n). If we allow ourselves the luxury of quantificational notation, we can see why this is so. In this notation,

$$X..LMY = (\exists z)(XLz \ \&\ zMY)$$
$$X..LM'Y = (z)(zMY \supset XLz)$$
$$X..L_{,}MY = (z)(XLz \supset zMY).$$

This suggests that there will be logical relations generated within De Morgan's system, since "X. .LMY" is a quantified statement, just as the other two are. The novelty comes from taking quantity, which is normally attached to the terms, and building it into the relation itself. This is a very deep feature of the system, for the validity of many proofs requires that LM, LM′, and $L_{,}$M be treated as relations, on a par with L and M.

De Morgan realized that the relationships between the notation for quantified relations and his usual notation for quantity is rather complex. For instance, in "LX))MY," the quantity is attached to the terms, yet it is equivalent to a proposition in which the relation is one of inherent quantity:

$$\text{LX))MY} = (z)(z\text{LX} \supset z\text{MY}) = (z)(\text{XL}^{-1}z \supset z\text{MY})$$

$$= \text{X. .L}^{-1}{}_{,}\text{MY.}$$

Thus, "Every parent of John is a friend of Mary" is equivalent to "John is a child of none but friends of Mary." We can well imagine the "confusions and blunders" which must have afflicted De Morgan in working out equivalences such as these (S4, 212).

One complexity to which De Morgan devoted little attention is the whole question of grouping. When a relation term contains more than two letters, some way of grouping the operations, such as parentheses, must be provided. Otherwise, syntactically ambiguous expressions can occur. For instance, "LMN′" could mean either "(LM)N′" or "L(MN)′", and these two expressions have different meanings.

De Morgan never groups in *S4*. There is very little reason to do so, for he seldom uses relation terms with more than two letters. On those few occasions when he does, the result is often not ambiguous–e.g., "$\text{lM}'\text{M}^{-1}$" and "$\text{L}^{-1}\text{L}_{,}\text{m}$". He also uses multiple compositions without parentheses, as in "LĹL," but composition is associative. One would have thought that De Morgan would have been alert to issues of grouping and associativity in logic because he was probably the first writer in English to see the need for such laws in algebra.[4]

As we will see in Chapter Eight, this duplication of notation between De Morgan's dot-notation and his spicular notation presents some serious problems of formalization. In a well-known article, Tarski has pointed out that the logic of relations can be developed in two different ways: within traditional quantification theory, utilizing both quantified individual variables and free relational variables, with axioms specific to the logic of relations; or, as a theory using only free relational variables, whose axioms state logical equivalences between complex relational expressions. These same two approaches are present in De Morgan's development of the logic of relations. The ideal seems to be a system in which only relational variables appear, but the justification for several laws often takes De Morgan back to their individual instances. Chapter Eight will discuss these options in greater detail, as well as a third option which is closer to recent work in the logic of relations.[5]

Unfortunately, the deduction of laws within this system is often obscure, since De Morgan does not provide an exact formulation of its basis. Intuitive arguments are used at times; quite formal ones, at other times. The principles of inference are never formally specified, so one must dig them out from the proofs themselves. However, De Morgan is able to establish many classic principles of the logic of relations, such as:

1. Not-L is the converse of not-(L^{-1}).
2. L^{-1} is the contrary of (not-L)$^{-1}$.
3. Not-(L^{-1}) is (not-L)$^{-1}$.
4. $M^{-1}L^{-1}$ is the converse of LM.
5. lM' and $L_{'}m$ are contraries of LM.
6. $m^{-1}L^{-1'}$ is the converse of the contrary of LM.

There are many other principles which relate combinations of relations and inherent quantities, but examples 5 and 6 will suffice to show the sophistication of the logical relations which De Morgan has detected.

The first four laws can be stated in English fairly easily:

1. The complement of a relation is the converse of the complement of its converse.
2. The converse of a relation is the complement of the converse of its complement.
3. The complement of the converse of a relation is the converse of its complement.
4. The composition of the converse of a relation with the converse of another relation is the converse of the composition of the second relation with the first.

Laws 5 and 6 seem almost impossible to state in English, since it lacks terms for the two forms of inherent quantity. We could, of course, define a term for the purpose, but the resulting English sentences would not be very helpful.

De Morgan's most formal and most interesting result is a theorem which he calls "Theorem K," in honor of its use in such syllogistic forms as Baroko (S4, 224–25). In this proof, he uses some unstated logical principles which can be discerned when the proof is unpacked. Theorem K has two parts, each of which implies the other (though De

Morgan did not comment on this equivalence). The proof of the first part requires three unstated rules of inference from the general theory of relations. They are:

R1. From L))M and M))N, infer L))N.
R2. If L||M then L may be replaced by M in any formula.
R3. From L))M, infer m))l.

Using these principles and sentential logic, the proof may now be given.

Theorem K: If LM))N then nM^{-1}))1.

Proof:		
	1. LM))N	Assumption
	2. n))not-(LM)	1, R3
	3. not-(LM)||lM′	Principle 5
	4. n))lM′	2,3, R2
	5. If n))lM′ then nM^{-1}))(lM′)M^{-1}	New Principle
	6. nM^{-1}))(lM′)M^{-1}	4,5, Modus Ponens
	7. (lM′)M^{-1}))1	New Principle
	8. nM^{-1}))1	6,7, R1

Theorem K is less mysterious than it looks. It merely shows that from "if X. .LY and Y. .MZ, then X. .NZ," one can infer "if X. .not-NZ and Z. .M^{-1}Y, then X. .not-LZ." In quantificational terms, the antecedent is,

$$(x)(y)[(\exists z)(xLz \,\&\, zMy) \supset xNy],$$

which is equivalent to,

$$(x)(y)(z)[(xLz \,\&\, zMy) \supset xNy];$$

and the consequent is,

$$(x)(y)[(\exists z)(\sim xNy \,\&\, yM^{-1}z) \supset \sim xLz],$$

which is just,

$$(x)(y)(z)[(\sim xNy \,\&\, yM^{-1}z) \supset \sim xLz].$$

De Morgan seems to have been intrigued by Theorem K, and especially by the following example of it (S, 186–87):

Every master of a parent is a superior.
Every servant of an inferior is a non-parent.
Every inferior of a child is a non-master.

If we assume that absolute equality is impossible (so that inferior is the contrary of superior), these three forms are equivalent, by Theorem K.[6] Another example, given by De Morgan for a somewhat different purpose, would be (S4, 225):

> Every brother of a parent of a person is an uncle of that person.

By the first half of Theorem K, this implies,

> No non-uncle of a child of a person is a brother of that person.

The proof of Theorem K does warrant a few comments. To begin with, R2 is not strictly needed for the proof, since Step 4 would follow from Steps 2 and 3 using R1, when it is recalled that L||M implies that L))M. Nevertheless, the use of replacement is so widespread in De Morgan's logic of relations that its employment seems natural at this point. The new principle in Step 5 is just an instance of the general principle,

R4. From L))M, infer LN))MN.

De Morgan does not state this principle in a general way, but it is surely an important principle in a logic which attempts to embody the combination of relations.

The new principle at Step 7 is an instance of "(LM′)M^{-1}))L." De Morgan's ability to see the necessity of this principle, and then to discern its intuitive soundness, is a testimonial to his logical skill. An example of the principle would be that if John loves every child of a parent of Mary, then John loves Mary.

There is a second half of Theorem K. It states that if LM))N, then L^{-1}n))m. The proof is similar to the proof of the first half, using different but related principles. At Step 5, the new principle is,

R5. If L))M, then NL))NM.

This is an interesting principle, for it is the relational analogue of the principle behind oblique inferences, where relational terms, "L" and "M," have replaced "man" and "animal." Corresponding to Step 7, there is the principle,

$$M^{-1}(M_{,}L)))L.$$

If John is a child of a parent of none but lovers of Mary, then John loves Mary.

De Morgan then notes that there does not exist an analogue of Theorem K for relational equivalence: x is an uncle of y just in case x is a brother of a parent of y, but it is not the case that x is a non-brother of y if and only if x is a non-uncle of a child of y. The proof for the analogue of Theorem K fails at Step 7, where inclusion in the other direction fails; in general, it is not the case that $L))(LM')M^{-1}$. From the fact that John is a brother of Mary, it does not follow that John is a (brother of every parent) of a child of Mary. (Parentheses have been added to disambiguate the English sentence.) Mary may not have any children at all; and if she does, she is not the only parent.

A property of equality is that if $x \times y = z$, then $x = z \times y^{-1}$. De Morgan notes that this property does not carry over to identity between relations: $LM||N$ does not imply that $L||NM^{-1}$. It is important to see De Morgan's diagnosis of what would go wrong in attempting to prove this property (S4, 225). The "proof" would be:

Derive:	$L\|\|NM^{-1}$	
Proof:	1. $LM\|\|N$	Assumption
	2. $(LM)M^{-1}\|\|NM^{-1}$	1, Analogue of R4
	3. $(LM)M^{-1}\|\|L(MM^{-1})$	Basic property of relative composition
	4. $L(MM^{-1})\|\|L$	Needed principle, but *false*
	5. $L\|\|NM^{-1}$	2, 3, 4,Transitivity, reflexivity of ||

To see what is wrong with Step 4, let us examine each half of it. One half is,

4a. $L(MM^{-1}))) L$.

This half is obviously incorrect: if John loves a parent of a child of Mary, it does not follow that John loves Mary, for John may love Harry, the other parent. According to De Morgan, the other half of Step 4 is true, so that we could use,

4b. $L))L(MM^{-1})$

to deduce,

5b. $L))NM^{-1}$.

It appears that De Morgan is wrong on this point. When John loves

Mary, it does not follow that John loves a parent of a child of Mary, for Mary may be childless. The inclusion holds only under the assumption that there is something which is a child of Mary.

De Morgan must have realized that there was a problem here, for he attempts to justify the claim that L))L(MM^{-1}), on the ground that $MM^{-1}X$ is a class which contains X. But this condition just means that $XMM^{-1}X$; that is, that ($\exists$y)(XMy & $yM^{-1}X$). This is the condition stated above, ($\exists$y)(XMy), that there is something of which Mary is a mother. Why, though, should De Morgan find this condition plausible? One explanation may be in De Morgan's doctrine of existential import, which assumes that all terms in categorical propositions are non-empty. If this is true in general, it might also be assumed to be true of a term defined relationally, such as "child of Mary." At the same time, one must be sceptical of any system in which this blanket assumption is made. This issue will be discussed again in Chapter Eight.

Having noted that $L||NM^{-1}$ does not follow from LM||N, De Morgan makes the more sweeping claim that this sort of thing can never arise. Given that LM||N, no identity can be deduced in which L is equivalent to any function of M and N, or in which M is identified with some function of L and N. It is interesting to note that De Morgan does not attempt to prove this theorem for the full resources of his logic of relations; he merely contents himself with a few simple constructions which obviously fail.

Before going on to more specific topics, we should note the viewpoint from which De Morgan developed his general logic of relations. He was primarily interested in doing two things: opening up the whole realm of relational inferences; and applying these relational forms to the traditional syllogism. He did not concern himself with many of the questions which a modern logician (or Aristotle, for that matter) would bring to the subject: the close analysis of deductive reasoning, the formal specification and completeness of the logic of relations, and its precise applicability to arguments in ordinary English. His concern was not with process or system but with results.

3. PROPERTIES OF RELATIONS

De Morgan then considers what have since become known as the standard properties of relations: convertibility, transitivity and, indirectly, reflexivity (S4, 225–27). A relation L is convertible when

X. .LY gives Y. .LX. For any relation L, $LL^{-1\prime}$ is convertible, and $L_{\prime}L^{-1}$ and $L'L^{-1}$ are converses. De Morgan then propounds a striking thesis: that every convertible relation is reflexive. (He does not use the word "reflexive.") The argument is in two stages, of which the first is:

> So far as I can see, every convertible relation can be reduced to the form LL^{-1}. If two notions stand in the same relation to one another, they can always, I think, be made to stand in one and the same relation to some third notion. The converse is certainly true, namely that two notions which stand in one relation to a third, stand in convertible relation to each other. But it cannot be proved that if X. .LY and Y. .LX, then L must be reducible to MM^{-1}, for some meaning or other of M: this is certainly a *material* proposition. But I can find no case in which material proof fails. (S4, 226)

In the second stage he argues:

> Among the subjects of a convertible relation must usually come the predicate itself, unless it be forced out by express convention. If all convertible relation can be expressed by LL^{-1} this is obviously necessary: for $LL^{-1}X$ includes X.

The argument, then, on which this claim is made has two premises:

Premise I: If L is a convertible relation, then there is some M such that $L \| MM^{-1}$.

Premise II: For any relation M, MM^{-1} is reflexive.

From these two premises it follows that any convertible L will also be reflexive. But are the premises true?

Let us begin with *Premise I*. It says that L is convertible if and only if,

$$D(L){:}\ (\exists M)(x)(y)[xLy \equiv (\exists z)(xMz \ \&\ yMz)].$$

As De Morgan points out, half of the claim is obvious: if condition D(L) holds, then L is convertible. In fact, this just derives the convertibility of L from the commutativity of "&." But is it true the other way around? Given any convertible L, can we always find such an M?

De Morgan admits that this is a "material proposition" which cannot be proved; he is merely unable to find a counterexample. Unfortunately, we do not know how hard he tried, since he gives a very limited set of examples. The most obvious type of positive case is that of the sibling relation, where x is a sibling of y if and only if there is some z such that x is a child of z and y is a child of z. (Here we count half-siblings as siblings.) Such relations as "classmate" and "countryman" are also of this type.

Notions which involve equality are also good candidates for this thesis. One might claim that x is the same length as y if and only if they both bear a common relationship to some real number which is the measure of their length; or, they bear the same relation to some standard object (e.g., the standard meter). In fact, in the case of equivalence relations such as equality of length, we can say that M is merely the membership relation to the class of all the elements which stand in the relation L to each other. Most simply, we can note that L, itself, can serve as the M for equivalence relations, since $L \| LL^{-1}$ whenever L is transitive, convertible and reflexive.

The problem is that not all convertible relations are of these types. Considering the positive integers as points on a line, we note that the adjacency relation (expressed by "next to") is convertible, but it seems not to involve a common relation to some other point. In fact, it is not reflexive. Thus, Premise I is not universally true.

This problem runs rather deep. Let us say that a relation is *weakly reflexive* if the following condition holds:

$$WR(L): (x)(y)[xLy \supset (xLx \ \& \ yLy)].$$

It can be shown easily that every relation which satisfies condition D is weakly reflexive. Thus any convertible relation, such as adjacency, which is not weakly reflexive cannot satisfy condition D, and it will serve as a counterexample to Premise I.[7]

Let us turn now to *Premise II*, the claim that MM^{-1} is always reflexive. De Morgan's examples had convinced him that this is always true, but other examples will show that it is not. Taking the universe of discourse to be people, let "P" stand for the parent relation and "C" stand for its converse, the child relation.

Then De Morgan is certainly correct about the composite relation CP. This has the form LL^{-1}, and it is certainly true that, in this universe, every x is a child of a parent of x (forgetting, for the moment about Adam and Eve).

The situation is quite different with PC, which is also of the form LL^{-1}. Little Max, a newborn, is not a parent of a child of Max, because he is not a parent at all. To say that $xLL^{-1}x$ is to say,

$$(\exists y)(xLy \ \& \ yL^{-1}x),$$

which is equivalent to,

$$(\exists y)xLy.$$

This is clearly a condition (we will later call it the Existence Condition) which some relations satisfy and others do not. Thus, Premise II is also false.

An interesting variant on this example is to contrast "x is a sibling of y," which means that x and y have a common parent, and "x is a co-parent with y," which means that x and y are parents of the same child. Both of these clearly satisfy Premise I; yet the first is reflexive and the second is not. They are both weakly reflexive.

De Morgan notices that it is sometimes strange to think of relations which satisfy Premises I and II as reflexive, but he justifies this on the ground that it is a harmless stipulation.

> Is a man his own brother? It is commonly not so held: but we cannot make a definition which shall by its own power exclude him, unless under a clause expressly framed for the purpose. . . . In common language the stipulation is or is not made, according to the casual presence or absence of the necessity for it. Put the question what relation to a man is his brother's brother, and most persons will answer, His brother: point out that the answer should be, Either his brother or himself, and a fair proportion will think that himself was included. I shall hold, for logical purposes, that the predicate *is* included among its own convertible relatives. (S4, 226)

This defense of the reflexivity of "brother" takes two forms. First, De Morgan notes that its reflexivity follows from the definition, using condition D, in terms of common parentage. The only way to rule out reflexivity would be to add a special condition whose only purpose would be to make the relation irreflexive. Thus, X would be the brother of Y just in case X and Y have the same parents *and* X is not identical with Y. This would ensure that "brother" is irreflexive: but it does not do so as a consequence of any deep fact about the relation, but only because its reflexivity has been ruled out by *fiat*.

To underscore the claim that this is just a stipulation, De Morgan says that it "is or is not made, according to the casual presence or absence of the necessity for it." It is a linguistic convention of no systematic import, sometimes invoked and sometimes not. Most people, he claims, will say that the brother of a brother is a brother, not noting that this means that a man is his own brother. When this is pointed out to them, a fair percentage will accept this result.

These two arguments are inconclusive at best. The fact that a specific non-identity clause is needed to rule out reflexivity does not mean that there is anything suspicious or merely stipulative about the clause. The question is whether this clause is needed to represent the way the relation is ordinarily understood. It should be noted, though, that if the clause is required, then condition D must be rejected, for a common relation to a common object would not suffice for convertibility. What does our ordinary language have to say on this matter? De Morgan's test is inconclusive. When I am asked whether a brother of a brother is a brother, my positive response may well be based on the assumption that this brother is a third person, so that I do not consider the case where the brother of a brother is that person himself. It is not at all clear that De Morgan is correct in saying that a "fair proportion" will agree that a man is his own brother when this is pointed out.

De Morgan obtains less problematic results when he turns his attention to transitive relations: that is, those L such that LL))L.[8] A multitude of logical relations come tumbling out from his previously established general laws when this condition is added. Thus, from LL))L there follow not only LLL))LL, and so on, but also such inclusions as:

$$L^{-1}L^{-1}))L^{-1}$$
$$L))l^{-1}l'$$
$$L^{-1}l))l$$
$$L^{-1}))L_{,}L^{-1}.$$

De Morgan gives verbal examples for each of these inclusions, using the transitive relation "ancestor"(S4, 227). He obtains such tongue-twisters as,

> An ancestor is always an ancestor of all descendants.
> A non-ancestor is always an ancestor of none but non-ancestors.
> Among non-descendants are contained all ancestors of non-descendants.

At this point, the logical power of De Morgan's methods is evident, for it is likely that his formalism has allowed him to discover and verify previously unknown relational inclusions.

De Morgan notes that,

> The mathematician forces the predicate itself among its own chain of successive relatives, whether the relation be transitive or not: x, as Φ^0x, appears in the sequence ... $\Phi^{-2}x$, $\Phi^{-1}x$, Φ^0x, Φ^1x, Φ^2x, ... There is a little tendency towards the same thing in

ordinary language, especially when the relation is transitive. Milton, in calling Eve 'the fairest of her daughters,' meaning female descendants in general, allowed $\Phi^0 x$ to be a case of $\Phi^n x$. Nothing but circumlocution avoids the same thing in our day, and by it language loses much force, or some precision. (S4, 227)

Once again the question arises whether we may stipulate that a relation is reflexive, for the purposes of logical ease or mathematical elegance. De Morgan's viewpoint is natural if we think of the matter functionally; but if we think of what the superscripts mean, it is less so. From a purely functional standpoint, to say that someone, y, is an ancestor of Mary is to say that there is some n, either positive or negative, such that $y = \Phi^n$Mary. It would then be plausible to include $n = 0$. What we really mean, though, is that ancestors of Mary are those y such that, for some negative n, $y = \Phi^n$Mary; and descendants of Mary are those y for whom there is some positive n such that $y = \Phi^n$Mary. From this standpoint, it is more plausible to say that Mary is neither an ancestor nor a descendant of herself. This is at least as plausible as saying that she is both an ancestor and a descendant of herself, which is what De Morgan seems forced to say.

It is puzzling to note that De Morgan does not develop as much as he might the simple condition for convertibility, that $L \| L^{-1}$. While he combines the transitivity condition, LL))L, with the relational laws which hold for all relations, he does not do the same for the convertibility condition; nor does he combine the convertibility and the transitivity conditions. Since so many of his relational laws involve formulas which contain both L and L^{-1}, the identity of L and L^{-1} can lead to striking results, directly applicable to the copulas used in his theory of the syllogism. These will be outlined in Chapter Eight.

With the discussion of convertible and transitive relations, De Morgan concludes his general theory of the logic of relations. He has outlined only enough of that theory to provide a foundation for treating the syllogism in relational terms, and it is to that discussion that we now turn.

4. SINGULAR RELATIONAL SYLLOGISMS (UNIT SYLLOGISMS)

We have already seen how the traditional syllogism can be treated by means of the composition of relations: since each premise of a traditional syllogism states a relation between two classes, the syllogism itself can be viewed as utilizing the composition of those relations.

The purpose of this section is to outline in a precise way the theory of the relational syllogism which lies behind this analysis and which can be generalized to any relations with the requisite properties. De Morgan's way of stating this analysis is by now familiar:

> The supreme law of syllogism of three terms, the law which governs every possible case, and to which every variety of expression must be brought before inference can be made, is this;–any relation of X to Y compounded with any relation of Y to Z gives a relation of X to Z . . . Any way of speaking of one notion with respect to a second, joined with a way of speaking of the second notion with respect to a third, must dictate a way of speaking of the first notion with respect to the third. And this is syllogism: it exhibits, in the most general form, the law of thought which connects two notions by their connexions with a third. The character of the connexions belongs as much to the *matter* of the syllogism as the character of the terms connected.
>
> The universal and all-containing form of syllogism is seen in the statement of X. .LMZ as the necessary consequence of X. .LY and Y. .MZ. Whether the compound relation LM be capable of presentation to thought under a form in which the components are lost in the compound–in the same manner, to use Hartley's simile, as the odours of the separate ingredients are not separately perceptible in the smell of the mixture–is entirely a question of matter. (S4, 227–28)

In interpreting these comments, it is important to remember the long-standing controversy over the priority of the first figure in the theory of the traditional syllogism. De Morgan gives to this one form of inference an absolute centrality in his theory of the "unit syllogism." While his "unit syllogisms" come in four figures and four phases (somewhat analogous to the traditional moods and figures), the validity of each is obtained in an obvious way from that of the primary relational inference. We shall attempt to understand De Morgan's method of categorization by discussing his classification of both figures and "phases."

a. Figures

In the traditional syllogism, the four figures depend upon the placement of the terms in the two premises. In De Morgan's scheme, it is a matter of the way in which the composition of relations enters (S4, 229–30). In De Morgan's terminology:

First Figure: direct transition

X. .LY
Y. .MZ
―――――
X. .LMZ

Second Figure: *reference to*

$$\begin{array}{l} \mathrm{X..LY} \\ \mathrm{Z..MY} \\ \hline \mathrm{X..LM^{-1}Z} \end{array}$$

Third Figure: *reference from*

$$\begin{array}{l} \mathrm{Y..LX} \\ \mathrm{Y..MZ} \\ \hline \mathrm{X..L^{-1}MZ} \end{array}$$

Fourth Figure: *inverted transition*

$$\begin{array}{l} \mathrm{Y..LX} \\ \mathrm{Z..MY} \\ \hline \mathrm{X..L^{-1}M^{-1}Z.} \end{array}$$

It is clear that the first figure is present in each of these figures, even if it is disguised. In the second figure, for instance, the fundamental syllogistic form is:

$$\begin{array}{l} \mathrm{X..LY} \\ \mathrm{Y..M^{-1}Z} \\ \hline \mathrm{X..LM^{-1}Z.} \end{array}$$

A full proof of the validity of the second figure form would first convert the second premise by converting its relation, thus bringing the entire inference into the standard first figure form.

So far, we have considered only affirmative premises and conclusions. When negative premises are considered, the full power of De Morgan's schematization can be seen. Consider, for instance, the premises X.LY and Z. .MY. From them we obtain,

$$\begin{array}{l} \mathrm{X..lY} \\ \mathrm{Y..M^{-1}Z} \\ \hline \mathrm{X..lM^{-1}Z.} \end{array}$$

We are now in a position to appreciate De Morgan's comment that "it was long before the *quantified relation* suggested itself, and until this suggestion arrived, all my efforts to make a scheme of syllogism were wholly unsuccessful." (S4, 221n) The problem is a simple one: the standard first figure form shows how to obtain an affirmative conclusion from two affirmative premises; but how can this affirmative conclusion be stated in a negative form? Only by having some way to negate compound relations. This requires De Morgan's quantified relations. Thus, the conclusion,

$$X. .lM^{-1}Z$$

can equally well be stated negatively as either,

$$X.LM^{-1\prime}Z \text{ or } X.l'm^{-1}Z.$$

The full logic of relations, utilizing contraries, converses, composition of relations, and both kinds of inherent quantity allows us to deal easily with any unit syllogism: using contraries and converses, the premises may be put into standard first figure form; this may then be used for the composition of relations to yield an affirmative conclusion; and this may be changed to a negative conclusion, by using one of the equivalences involving inherent quantity.

De Morgan points out that if contraries or inherent quantity were not available, we would be forced to use *reductio ad absurdam*. Let us consider again,

$$\begin{array}{l} X.LY \\ \underline{Z. .MY} \\ X.NZ, \end{array}$$

where N is temporarily unknown. The task is to find N. By the opponent syllogism, this is equivalent to,

$$\begin{array}{l} X. .NZ \\ \underline{Z. .MY} \\ X. .LY. \end{array}$$

These premises also yield, X. .NMY, so that L||NM. The problem, though, is how to move from this to the desired conclusion of the original argument–that is, the form which does not use contraries, which is $X.LM^{-1\prime}Z$. De Morgan first justifies this transition with an example: if we say that X is not an uncle of Y, and that Z is a parent of Y, then we know that X is not an uncle of a particular child of Z–namely, Y. If we then drop all reference to Y, we can conclude that X is not an uncle of all the children of Z (S4, 230). He later attempts to prove the desired conclusion in a more abstract way:

If we examine NM||L, upon the condition that M is some one M, left vague, but not any one M, we see that it gives N||LM^{-1}. For L is to include in its meaning any N of a certain M, and nothing else: so that N is LM^{-1}, where that M^{-1} is used which is the correlative of the M in question. But in *denying* this LM^{-1}, or rather any L of this vague M^{-1}, we do but deny LM^{-1}'. (S4, 223)

The line of reasoning here is obscure. We have already seen that, in general, NM||L does not imply N||LM^{-1}. What De Morgan appears to argue is that from the two original premises, given that NM||L, we may infer that we are denying that X is an L of some particular M^{-1} of Z–namely, Y. But we may drop the reference to the particular Y, which means that we are denying that X is an L of all M^{-1} of Z.

Having introduced the full system of contraries and inherent quantities, it is odd that De Morgan reverts to the method of relational decomposition which tried to avoid them. We shall see, though, that he uses both methods in outlining his full system of unit syllogisms.

b. Phases

The usual theory of the syllogism uses both figures and moods, specifying for the conclusion and each premise whether it is an A, E, I or O proposition. De Morgan has no need of this classification, for he has only affirmative and negative propositions; furthermore, the standard conclusion may be taken to be the affirmative conclusion generated by the composition of relations. Rather than the many moods of the traditional syllogism, then, De Morgan has just four possibilities for combinations of relations (S4, 231–32). With "+" for affirmative and "–" for negative, these are:

$$+\ +,\quad -\ +,\quad +\ -,\ \text{and}\ -\ -.$$

The simplest system would be to assign the numbers 1 through 4 to these four possibilities. In fact, De Morgan uses a much different scheme in which the phase number of a given configuration depends not only on the distribution of + and –, but on the figure as well. This is as follows:

		Phase 1	2	3	4
Figure	I	+ +	– +	+ –	– –
	II	– +	+ +	+ –	– –
	III	+ –	– +	+ +	– –
	IV	– –	+ –	– +	+ +

Unfortunately, De Morgan gives no justification for this odd arrangement. He does call the first phase of each figure its *primary phase*, suggesting that its distribution of + and − is most characteristic of that figure. Given the primary phase, the ordering of each phase for each figure is based on a rule which seems calculated more to help one remember a unique order than to illuminate the logic of the unit syllogism. At this point, there seems to be little reason for De Morgan's peculiar arrangement of phases; its rationale is to be found only in its application to the traditional syllogism.

De Morgan tabulates these sixteen forms of syllogism in boxes, each of which contains six lines (S4, 232–34). Three examples will show what is at issue.

I.1	*II.1*	*III.4*
$\mathrm{X.\ .LY}$	$\mathrm{X.LY}$	$\mathrm{Y.LX}$
$\mathrm{Y.\ .MZ}$	$\mathrm{Z.\ .MY}$	$\mathrm{Y.MZ}$
$\mathrm{X.\ .LMZ}$	$\mathrm{X.\ .lM^{-1}Z}$	$\mathrm{X.\ .l^{-1}mZ}$
$\mathrm{X.lM'Z}$	$\mathrm{X.LM^{-1\prime}Z}$	$\mathrm{X.L^{-1}m'Z}$
$\mathrm{X.L_{,}mZ}$	$\mathrm{X.l_{,}m^{-1}Z}$	$\mathrm{X.l_{,}{}^{1}MZ}$
$\mathrm{LM\Vert N}$	$\mathrm{NM\Vert L}$	$\mathrm{l^{-1}m\Vert N}$

The first two lines in each example are the premises, while the third line is the conclusion obtained by reducing the form to the primary phase of the first figure. The fourth and fifth lines are the negation equivalents of the third line, obtained by using the standard relational equivalences for quantified relations. The sixth line requires some explanation:

> In the sixth line N is the concluding relation, affirmative or negative according as the premises are of similar or different qualities: and the connexion of N with the premising relations is seen, as obtained by simple composition when the premises are of the same quality, and from opponent syllogism, or from the rule above, when the premises are of different quality. (S4, 233)

The meaning of this quotation can best be seen through the three examples. I.1 is the primary phase of the first figure, so that its third line shows the composition of relations applied to the first two lines. In lines four and five, the relations of inherent quantity are the standard contraries of LM. The concluding relation, N, is just the composition of L and M.

Example III.4 contains two negative premises, and it can easily be reduced to I.1 by changing its form to,

$$\begin{array}{l} X..l^{-1}Y \\ Y..mZ \\ \hline X..l^{-1}mZ. \end{array}$$

Lines four and five are obtained by standard equivalences. In explaining the sixth line, De Morgan says that since "the premises are of one quality, so that N is not to be disengaged by decomposition, it is enough that N should be identical with, or should contain, the relation set down before it." Thus, any inference from these premises to a conclusion of the form X. .NZ will be valid, as long as N includes $1^{-1}m$ (which allows that N could be identical with $1^{-1}m$).

We have now considered cases in which both premises are affirmative and both are negative. Example II.1 is a case in which one premise is affirmative and the other is negative. This form is equivalent to,

$$\begin{array}{l} X..lY \\ Y..M^{-1}Z \\ \hline X..lM^{-1}Z. \end{array}$$

The fourth and fifth lines are obtained from this by standard equivalences. The sixth line, though, is determined in a way different from that used when both premises are of the same quality. The method of decomposition is used here; in fact, it is just the case which we considered in our discussion of figure, where the *reductio* method showed that NM||L. The informal, but problematic, discussion therein provided an alternative derivation of line four, rather than using the standard contrary of line three.

It should be noted that N stands for the relation which is *affirmed* when both premises are of the same quality, and for the relation which is *denied* when they differ in quality. This distinction is unnecessary, for the method of decomposition is not required when contraries of relations are allowed. It seems that much of this puzzling schema is dictated by the notion that in syllogisms with premises of the same quality, the ordinary form of the conclusion is affirmative; while, when they differ, it is negative. Given the ease with which De Morgan is able to move from a relation to its contrary, this seems like an unimportant distinction.

5. QUANTIFIED RELATIONAL SYLLOGISMS

We have seen how, in *FL*, De Morgan suggested that quantified syllogisms be validated by using the properties of relations found in their instances. This is now possible for the full logic of relations, though De Morgan only sketches how it might be done. His purpose is to develop only enough of that theory to allow a quantified relational treatment of the syllogism. This treatment is not absolutely necessary, since the theory of unit syllogisms suffices for the traditional syllogism. The traditional categorical propositions can be construed as stating relations between classes, from which the conclusion arises through the combination of relations. In this method, quantification has been thrown into the relation:

> the quantified syllogism of relation is not of primary importance as an explanation of actual thought: for by the time we arrive at the consideration of relation in general we are clear of all necessity for quantification. And for this reason: quantification itself only expresses a relation. Thus if we say that some Xs are connected with Ys, the relation of the class X to the class Y is that of *partial connexion*: that some at least, all it may be, are connected, is itself a connexion between the *classes*. Nevertheless, it may be useful to exhibit the modifying quantification as a component, not as inseparably thought of in the compound . . . (S4, 234)

De Morgan rightly notes that when terms, quantifications and relations are allowed to mix, a wide variety of new forms emerge: we would want to be able to deal with such forms as:

All L of some X are M of some Y.
Some L of all Y are M of some Z.
No L of some X are M of all Z.

This would open up whole new fields of logic where, "as in mathematics, the horizon opens with the height gained: generalization suggests detail, which again suggests generalization, and so on *ad infinitum*"(S4, 235). De Morgan is not in a position to develop this generalized logic, so that he confines his attention to the "Aristotelian" branch.

Before seeing how this branch can be developed, let us examine De Morgan's claim that this extension is not really necessary, that it is contained in the logic of the unit syllogism which he has already outlined. In a sense this is true. Each of the three complex statements above asserts a relation between the classes X and Y, even though there might be no common name for this relation. It would be possible to construct arguments using forms such as these, with one premise asserting a complex relation between X and Y, and the other premise asserting a complex relation between Y and Z. The conclusion would be that X stands in the compound relation to Z. The problem, of course, is that there will be different relations for every possible form, of which there are a great many; and, if there are n relations, there will be n^2 different combinations of relations. Furthermore, there will be no general principles for reducing compound relations to single relations. In the traditional 1 AAA syllogism, for instance, the initial form of the conclusion is X. .LLY; but, since LL))L, one may conclude that X. .LY. Since the relations involved may be any relations whatever, the reduction principles will be completely specific to them and thus material principles.

This general point is somewhat obscured in De Morgan's treatment. By confining his discussion largely to the traditional syllogism, with its limited number of logical forms, the potential for logical chaos is reduced. But Chapter Six will show us that even this case is not as simple as it looks, for the number of compound relations in De Morgan's version of the traditional syllogism is quite high.

It seems that quantified syllogisms can be treated more perspicuously by confining the unit syllogism to propositions with true singular terms, and then reducing quantified syllogisms to them through instantiation and generalization. This, in effect, is what De Morgan does in showing how the quantified forms can be used. We shall see that this may be preferable even in the simple case of the traditional syllogism.

Several questions must be answered in moving to the quantified syllogism. The first question is purely notational. De Morgan uses "X" and "Y" to stand both for the singular terms in unit syllogisms and the class terms in quantified syllogisms. This usually presents no problems, especially since the terms in unit syllogisms which correspond to the traditional syllogism are themselves class terms. But when we are reducing quantified syllogisms to unit syllogisms, there is the potential for confusion. For the time being, then, lower case letters from the beginning of the alphabet will be used for the terms in unit syllogisms; and upper case letters, for terms in quantified syllogisms. As usual, the complement of a class term will be expressed by the corresponding lower case letter–e.g., "x" stands for non-X.

The next question to be answered concerns the scope of allowable propositional forms. On the propositional level, should relations be allowed in both subject and predicate, or only in one? While in theory it seems that they could be allowed in both, De Morgan confines his attention to cases in which a relation appears in just one term. Similarly, what about the syllogism? In the first figure of the quantified syllogism, must the predicate of the first premise be the same as the subject of the second, or need they only contain the same class or singular term? De Morgan adopts the latter alternative for the restricted purposes of his discussion, though it seems there could be many options.

Finally, when "X" is a class term, what does "LX" stand for? Does it stand for the class of things which stand in the relation L to some X or other, or for those things which stand in L to all Xs? In generalizing from the unit syllogism, we could select either option, since these options coincide for singular terms. De Morgan selects the first option, thus making relation-class composition, LX, similar to relation-relation composition, LM. It is also the mode of composition in oblique inferences.

For the very limited class of cases which De Morgan has in mind, some typical syllogisms for class inclusions will be:

Ex.1	*Ex.2*	*Ex.3*
X)) LY	X).(LX	X)) LY
Y)) MZ	Y((MZ	MY).)Z
X)) LMZ	X).(LM′Z	$X).)L_{,}M^{-1}Z$

Let us now see how each of these forms can be proved through reduction to unit syllogisms.

Ex.1

Let a be a member of X. Then there is some member of Y, call it b, such that a. .Lb. But then, by the second premise, there will be some member of Z, call it c, such that b. .Mc. We now have the unit syllogism,

a. .Lb
b. .Mc
———
a. .LMc.

But, the conclusion means that there is some c in Z to which a stands in the relation LM, so that a. .LMZ. And, since this holds for any a in X, X))LMZ.

Ex.2

This case is more complex. If a is in X, then for every b in Y, a.Lb. Consider some b in MZ. Then there is a c in Z, such that b. .Mc; but by the second premise, this b is also in Y, so that it falls under the first premise. Thus we have,

a.Lb
b. .Mc
―――――――――
a. .lMc, which is a.LM′c.

This, says De Morgan, allows us to move to X).(LM′Z. In fact, this move is illicit, unless strong existential assumptions are made. In order to justify this move, it must be assumed that the c in question is an arbitrary member of Z; but it is not, for it is only a Z which makes it true that b is an MZ, for arbitrary b.

Another way to see the same point is to compare the original conclusion of this argument with that of Ex.1. In Ex.1, there is the valid inference from a. .LMc to a. .LMZ: if a is an L of an M of c, and c is a Z, then a is an L of an M of a Z. Similarly, in Ex.2, if a is an lM of c, and c is a Z, then a is an lM of a Z. However, a. .lMZ does not imply a.LM′Z, even though a. .lMc is equivalent to a.LM′c. This can be seen by using quantificational notation:

$$\text{a. .lMZ} = (\exists w)[Zw \;\&\; (\exists v)(alv \;\&\; vMw)]$$

$$\text{a.LM}'\text{Z} = \sim(\exists w)[Zw \;\&\; (v)(vMw \supset aLv)].$$

In going from assertion to denial, when both subject and predicate are singular terms, one merely changes the relation to its contrary; but this does not happen when the predicate is a class term.

This same phenomenon can be seen in the fully quantified version. Having concluded a. .lMZ, we could infer X))lMZ, since the assertion held for arbitrary a in X. But we cannot then use the fact that LM′ is a contrary of lM to argue that X).(LM′Z. In general, we may argue from X))Y to X).(y. But LM′Z is not the contrary of lMZ, even though LM′ is the contrary of lM. This point is the basic one, for it holds for all relations, not just complex ones. For any L,

$$X))LZ = (v)[Xv \supset (\exists w)(Zw \;\&\; vLw)] \text{ and}$$
$$X).(lZ = (v)[Xv \supset \sim(\exists w)(Zw \;\&\; vlw)]$$
$$= (v)[Xv \supset (w)(Zw \supset vLw)].$$

These two expressions are not equivalent. De Morgan needs a way to express the complement of a relation-class term, so that he can show that X))LZ if and only if X).(non-(LZ). For this, though, he needs some additional notation. To find out what this is, let us express X).(non-(LZ) in quantificational notation:

$$X).(\text{non-}(LZ) = (v)[Xv \supset \sim(\exists w)(Zw \;\&\; vLw)]$$
$$= (v)[Xv \supset \sim(w)(Zw \supset \sim vLw)]$$
$$= (v)[Xv \supset \sim(w)(Zw \supset vlw)].$$

That is, all X are an L of a Z if and only if no X is an l of every Z. To express this, we need a second means of joining L and X, to mean L of every X.[9] Using "*" for this purpose, we can use "l*Z" to express non-(LZ). The sentence in the example above will be expressed by X).(l*Z.

It should be noted that this new notation will complete the similarity between relation-relation composition and relation-class composition. There is ordinary composition in each case: LM and LX. We have now seen that we need the first kind of inherent quantity in both cases: LM′ and L*X. The second kind of inherent quantity can also be defined for relation-class composition, so that we have L of none but X, just as we have $L_{,}M$. While De Morgan admits the notion of L of every X, just as much as L of some X, he utilizes only the latter in developing his theory of the quantified syllogism. We have now seen that this will not do.

Let us return to the example which prompted this extension of De Morgan's system. We obtained X))lMZ. Then X).(non-(lMZ) will be X).(non-(lM)*Z, i.e., X).(LM′*Z, where "*" stands between the relation LM' and the class Z. This checks out quantificationally, since,

$$X).(LM'{*}Z = (v)[Xv \supset \sim(w)(Zw \supset (t)(tMw \supset vLt))]$$
$$= (v)[Xv \supset (\exists w)(Zw \;\&\; \sim(t)(tMw \supset vLt))]$$
$$= (v)[Xv \supset (\exists w)(Zw \;\&\; (\exists t)(tMw \;\&\; vlt))]$$
$$= X))lMZ.$$

De Morgan's error consists in writing LM′ of some Z when he should have written LM′ of every Z. He would have needed an additional notation for this. If he had had this notation, though, the rest of the inference, using reduction to the unit syllogism, would have been simple. Having inferred that a.LM′c, for some c in Z, one can conclude that a is not an LM′ of every c in Z. And, since this is true for an arbitrary a in X, one has X).(LM′*Z.

Ex.3

De Morgan's third example also shows the hazard of appeals to intuition, even by someone of De Morgan's caliber, though the lessons to be learned are less interesting (S4, 236). He attempts to deduce a conclusion from X)) LY and MY)(Z by transforming the latter premise into MY).)z and then to Y).)M^{-1}z to yield,

X)) LY
Y).)M^{-1}z
———————
X).)L,M^{-1}z.

Unfortunately, this argument is not valid. To see this more clearly, let us neglect the use of the converse and contrary, and consider another argument with the same basic form:

X))LY	All X are an L of some Y.
Y).)MZ	Some M of a Z are not Y.
X).)L,MZ	Some L of none but M of a Z are not X.

The problem with this argument is best seen by considering its equivalent opponent form:

L,MZ))X	All L of none but M of a Z are X.
X))LY	All X are an L of a Y.
MZ))Y	All M of a Z are Y.

The validity of this argument is made plausible by noting that the premises do imply L,MZ))LY. What must now be asked is whether,

L,MZ))LY	All L of none but M of a Z are L of a Y.
MZ)) Y	All M of a Z are Y.,

is valid. An example of this inference sounds rather convincing:

All lovers of none but servants of a Greek are lovers of a philosopher.

All servants of a Greek are philosophers.

The premise is ambiguous; but no matter how it is disambiguated, the inference is invalid.

The first reading is,

(v)(if there is one Greek of whom everyone whom v loves is a servant, then v loves a philosopher).

(w)(if w is a servant of a Greek, then w is a philosopher).

This argument is plainly invalid, for there may be many unloved servants of Greeks, and they need not be philosophers. The second reading would translate this premise as,

(v)(if everyone whom v loves is a servant of some Greek or other, then v loves a philosopher).

The argument using this premise also fails, and for the same reason.

Expressing this ambiguity provides another reason for introducing a "none but" operator into relation-class composition. "v is a lover of none but servants of a Greek" can mean either:

v is a (lover of none but servants) of a Greek
$(\exists w)[Gw \,\&\, (t)(vLt \supset tSw)]$

or,

v is a lover of (none but servants of a Greek)
$(t)[vLt \supset (\exists w)(Gw \,\&\, tSw)]$.

The first case can be expressed in De Morgan's notation as v. .L,SG, but the second requires a new notation to express "none but" when it occurs between a relation term and a class term. Using "#" for this purpose, we have, v. .L#SG.

De Morgan's two errors show that the formulation of a full theory of the quantified syllogism is a complex matter. De Morgan himself did not attempt to do so. There are no indications in De Morgan's published and unpublished papers and letters that he ever attempted to pursue the logic of relations any further. He did not touch the general logic of relations or the logic of quantified relational syllogisms again.

De Morgan never explained why he did not develop the logic of relations more fully. Perhaps the sheer complexity of the task led him to abandon it: De Morgan apparently did not have the tenacity we associate with the very finest mathematicians. Undoubtedly, he also was limited by his concentration on that part of the logic of relations which is needed for the theory of the syllogism. The formulation of an extended logic of relations would have to wait some ten years until Peirce's classic memoir of 1870 on the logic of relatives.[10]

6. THE LIMITED UNIT SYLLOGISM

De Morgan consistently held that the traditional syllogism is merely a special case of the relational syllogism. This is shown in *S4* when De Morgan returns to the unit syllogism and confines himself to unit syllogisms in which:

a) Only one relation and its converse appear in the premises; and,
b) That relation is transitive. (S4, 236–37)

Then either L or L^{-1} will be the relation in the premises, and both LL))L and $L^{-1}L^{-1}$))L^{-1}. Since there are a vast number of inclusions for transitive relations, many of the conclusions of these limited unit syllogisms imply propositions using only L or L^{-1}. The three examples of unit syllogisms we used can illustrate this.

Example I.1 becomes,

X. .LY
Y. .LZ
———
X. .LLZ
———
X. .LZ.

The last step follows, of course, since LL))L. Applying II.1, we obtain,

X.LY
Z. .LY
———
X.$LL^{-1\prime}$Z
———
X.LZ.

The final step is based on the fact that, for transitive L, L))$LL^{-1\prime}$. Finally, Ex.III.4 is like the first three figures of the fourth phase in not having a conclusion which uses only L or L^{-1}. It can give,

$$\frac{\begin{array}{l} Y.LX \\ Y.LZ \end{array}}{X..ll^{-1}Z,}$$

for which there is no appropriate inclusion using only L or L^{-1}. The same holds for the negated forms of the conclusion, $X.L^{-1}l'Z$ and $X.l_{,}^{-1}LZ$.

Actually, each figure and phase can have two possible syllogisms. If, to standardize matters, the first premise is allowed to contain only L, then the second premise can use either L or L^{-1}. So far we have considered only the first alternative in discussing III.4. The second alternative is,

$$\frac{\begin{array}{l} Y.LX \\ Y.L^{-1}Z \end{array}}{X..l^{-1}l^{-1}Z,}$$

which, again, gives no simple conclusion in just L or L^{-1}. The same is true of the negative forms of this conclusion.

De Morgan emphasizes that these complex conclusions are still legitimate conclusions: it is just that they are not "complete analogues" of the common syllogism. They may even be important. The other form of I.1 would be,

$$\frac{\begin{array}{l} X..LY \\ Y..L^{-1}Z \end{array}}{X..LL^{-1}Z,}$$

which cannot be further reduced. He comments that "still it is a conclusion, and to some persons an important one: for if L means *descendant*, and therefore L^{-1} *ancestor*, then, Z being the Queen, X is entitled to an honorary degree" (S4, 237).

In looking at these limited forms of unit syllogisms, we can discern the probable rationale for De Morgan's peculiar system of figures and phases. They are set out in such a way that the results can be reasonably applied to unit syllogisms with a single transitive relation and its converse. The first phase forms for the first three figures then become the "radical" or "primitive" forms in which the converse is mentioned neither in the premises nor in the conclusion. The fourth figure has no phase without some use of the converse. De Morgan also notes that when just one premise differs from that in the "radical" form, the other premise must use the contrary; and that when both premises differ in this way, the conclusion uses the converse.

This may not completely explain the set-up for the unit syllogisms, but it helps. The general schema can be seen more clearly if we let letters stand for each of the four combinations, in their standard order:

$$A = +\ + \quad B = -\ + \quad C = +\ - \quad D = -\ -$$

		Phase 1	2	3	4
Figure	I	A	B	C	D
	II	B	A	C	D
	III	C	B	A	D
	IV	D	C	B	A

The sequence of letters for each figure is determined by placing the primary phase in the first phase, then working to the left in the standard order (if there are any) and then working to the right (if there are any). The result is that all the + + forms are on the diagonal, though it is not clear that there is any systematic importance to this fact.

7. THE ORDINARY SYLLOGISM AND THE RELATIONAL SYLLOGISM

De Morgan concludes this important memoir by comparing the roles of the ordinary syllogism and the relational syllogism–both in "ordinary" reasoning and in the sophisticated reasoning of mathematics. We will deal at length with his views on mathematical reasoning in Chapter Seven, so that at this point we will confine our attention to everyday forms of reasoning.

According to De Morgan, the ordinary syllogism is a valid, but comparatively rare, feature of ordinary reasoning; furthermore, when it does occur, it is usually in the simple forms of *Barbara* and *Celarent*. The other traditional features of inference are of more significance.

> Of the common logical heads, the study of the term and of the proposition, of the aggregates and components of the term, and of the transformations of the proposition, is far more necessary, presents points of far more frequent occurrence, and holds out far greater occasion for warning, than the study of the syllogism . . . (S4, 237)

The traditional syllogism, then, usually occurs in a simple form, and the interesting material is elsewhere. Of more interest is the study of logical relations between complex terms, constructed from simpler terms by contrariety, conjunction, and disjunction. Of more interest, too, are the logical relations between single propositions, especially those "transformations" by which single propositions are inferred from single propositions. This can include such cases as contraposition, instances of De Morgan's laws, and oblique inferences. The most interesting inferences are required in order to put propositions into a form where the syllogism can be applied to them.

A second reason for the rarity of the ordinary syllogism is that it is confined to onymatic relations. De Morgan has emphasized repeatedly that the syllogistic forms can also be exhibited by many other relations as well. The prevalence of this form of reasoning can be seen by noting that in the process of discovery, we often find ourselves looking for a middle term for a relational syllogism, while we seldom do this for an ordinary syllogism. Traditional logic explicitly covered the means by which middle terms could be found, so that a given conclusion could be demonstrated. The disuse into which this aspect of traditional logic has fallen shows the lack of importance of the traditional syllogism. In granting ourselves the full range of relations, though, we can see just how common and important this type of investigation is. In trying to show that Z is a consequence of X, we look for some Y which is a consequence of X, and of which Z is a consequence. A complex instance of this would be our attempt to find a motive Y, so that we can argue,

> X is the possessor of the motive Y.
> Y is a sufficient motive to commit Z.
> ———————————————————
> X is the possessor of a sufficient motive to commit Z.

From this connection between X and Z, we can then infer,

> X is sufficiently in connexion with the action to give the evidence of actual commission of Z a claim to consideration under ordinary notions of probability.

De Morgan concludes,

> The old logicians were right in attaching importance to the invention of the middle term; but their right notion was deprived of efficient action by their determining that no

connexion worthy of a logician's attention could exist between terms except *is* or *is not.* (S4, 238–39)

It is unlikely that any two arbitrary notions will be related by any onymatic relation worth considering; but other relations between them will exist, which the logic of relations allows us to handle.

De Morgan's conclusion, then, is this:

When we examine any book of ordinary reasoning, we find that the onymatic syllogism is not very frequent, the combination of relations much more frequent, and the introduction of composition of terms and transformation of propositions by far the most frequent of all. Syllogisms are *rather chapters than sentences*, in many cases. . . . Nothing that I know of can be written *all* in syllogism, except mathematics: and this merely because, out of mathematics, nearly all the writing is spent in loading the syllogism, and very little in firing it. (S4, 239)

Aside from questions of frequency, then, De Morgan is maintaining here that ordinary reasoning exhibits four kinds of reasoning:

Onymatic syllogism (traditional syllogism)
Combination of relations (the "unit syllogism")
Composition of terms (similar to the Boolean algebra of terms)
Transformation of propositions (traditional transformations, such as contraposition, along with relational transformations, including both the bulk of the logic of relations and oblique inferences).

These same four components of reasoning will be found in De Morgan's discussion of the logic of mathematics, though with differing frequencies. This seems to represent, then, De Morgan's settled view about the resources needed to represent all deductive reasoning.

We may naturally ask how adequate this account is. Would it be possible to exhibit some reasonably complex examples of deductive reasoning, either in ordinary life or in mathematics, in such a way that it is rigorously formulated and uses only these four types of principles? This is a difficult question to answer, and De Morgan does not really address it. While he has given a rather precise account of the onymatic and relational syllogisms, the application of the relational syllogism to specific relations depends upon additional postulates concerning their formal properties and the results of combining relations in various ways. The third and fourth classes of inferences are not well defined, for De Morgan did not construct a precise axiomatization of his system. However, De Morgan did give a great many forms of inference from which some idea of the scope of De Morgan's logic may be inferred. A clearer understanding of this scope will be one point of Chapter Eight.

We have now had the chance to examine not only De Morgan's general logic of relations, but many of its applications as well. Questions of application loom largest when we consider the details of De Morgan's relational treatment of the traditional syllogism; and it is to this subject that we now turn.

THE LOGIC OF RELATIONS AND THE THEORY OF THE SYLLOGISM

Writers on De Morgan have often distinguished between the two most important facets of his work: his theory of the syllogism and his logic of relations. The former looks to the past; the latter, to the future.

While there is some truth to this interpretation, it should be qualified in two ways. De Morgan's analysis of the syllogism is shaped by his logic of relations, for he construes the theory of the syllogism in terms of the logic of relations. Conversely, his logic of relations is developed with the theory of the syllogism in mind. It is developed only to the point where the theory of the syllogism can be construed within it.

The purpose of this chapter is to determine precisely the relationship between these two aspects of De Morgan's work. Unfortunately, this is not an easy task. The problem lies not so much with the logic of relations as with the theory of the syllogism. De Morgan's logical writings are a succession of attempts to re-state and expand the theory of the syllogism. Both the symbolism and the interpretation change in the course of time. Furthermore, these systems are of considerable complexity. While they are not conceptually difficult, the variety of symbolisms and interpretations creates a serious expository problem.

To deal with this problem, we will concentrate on De Morgan's *Syllabus of a Proposed System of Logic*, dated 1860, which provides the most systematic and complete outline of his mature views (S, 147–207). Within the *Syllabus*, we shall concentrate on the most central features of his theory of the syllogism, and on the manner in which he relates them to the logic of relations.

De Morgan divides the *Syllabus* into two parts, the "Objective View" and the "Subjective View." We will do the same, for these represent two quite different ways of approaching the syllogism. Interestingly enough, they also embody two very different ways of applying the logic of relations to the theory of the syllogism.

We shall first consider one form of the objective view. On this account, each syllogism is viewed as the collection of all the unit syllogisms which fall under it. For instance, first figure AAA is the collection of the forms, "This X is a Y; that same Y is a Z; therefore, this X is a Z." To validate this syllogism, only the transitivity of identity is needed; in other forms, convertibility is needed as well. The doctrine of the abstract copula is central to this technique.

In sections two and three, we show how De Morgan provides a formal characterization of this approach to the syllogism. He develops the full syllogistic theory using "is joined to" as his transitive and convertible copula. The formal effect of dropping the convertibility condition is then shown with the relation "gives to." In the absence of convertibility, some way must be found to express the converse of a relation, and it will be seen that problems arise when De Morgan tries to combine this with his standard "spicular" notation for the syllogism.

De Morgan's second way of formulating the syllogism, to be outlined in section four, is much simpler. This is what De Morgan calls the mathematical subjective interpretation, and it reflects the bicopular syllogism and the composition of relations. From this standpoint, each of De Morgan's eight propositional forms states a relation between classes, and a traditional syllogism just involves the composition of these two relations. The resulting twenty-four regular syllogistic forms can be broken down into eight groups of three relational inclusions, where two members of each group are obtained from the other by using Theorem K. Alternatively, the whole system can be exhibited in a multiplication table which gives the result of "multiplying" the two relations in the premises.

1. THE TWO VIEWS

The distinction between objective and subjective views derives from the two ways in which names can be used, roughly corresponding to the old distinction between terms of first intention and terms of second intention (S, 178–79). An *objective* use of a term is directed toward the external object. This, in turn, can be either of two kinds:

1. The *individual object*, "unconnected with, and unaggregated with, any other object of the same name"; and,
2. The *individual quality*, "forming part of, and residing in, the individual object."

When a name is used in a *subjective* sense, it is "directed towards the *internal* mind, of *second intention*, or representing *second notions*."

This can also be in one of two ways:

3. A *class*, "a collection of individual objects, named after a quality which is in thought as being in each one"; and,
4. An *attribute*, "the notion of quality as it exists in the mind to be given to a class."

Related names give us these four types of terms: that is, "the objects *men*, all having the quality *human*, give to the thought the class *mankind*, distinguished from other classes by the attribute *humanity*" (S, 179). De Morgan suggests using the four terms "X," "X-ic," "X-kind," and "X-ity" to stand for an arbitrary term of each type.

We need not pursue here the obvious metaphysical issues behind these distinctions. Suffice it to say that these four kinds of terms provide several ways of saying the same thing. When we discuss the subjective view, we will even see that there are four different "subjective" ways of making common assertions. For the moment let us note that the following two statements, in some sense, "enunciate" the same proposition:

Every man is an animal.	(*Objective*)
The *class* man is contained in the *class* animal.	(*Subjective*)

It appears that the first of these statements asserts something about all individual men, while the second asserts a relation between two classes. De Morgan takes this apparent difference very seriously, and he builds two quite different theories of the syllogism on their bases. The logic of relations enters into both of these two methods, but in very different ways.

2. OBJECTIVE VIEW–THE BASIC ACCOUNT

The common separation between De Morgan's theory of the syllogism and his logic of relations is made plausible by the fact that De Morgan first develops the former without reference to the latter. The basic elements in the study of logic are defined as follows:

1. The *term* or *name*, the written or spoken sign of an object of thought, or of a mode of thinking.
2. The *copula* or *relation*, the connexion under which terms are thought of together.

3. The *proposition*, terms in relation with one another; and the *judgment*, the decision of the mind upon a proposition: usually joined in one, under one or other of the names.
4. The *syllogism*, deduction of relation by combination of other relations. (S, 153–54)

While this conception of logic is fully relational, De Morgan immediately restricts his attention to the special case of what he calls *onymatic* relations. From this standpoint, a proposition is "no more than the connexion of name with name, as marks of the same object," so that "Every X is Y" merely asserts that every object which has the name "X" also has the name "Y"; that is, "Y" applies to every object to which "X" applies. This *arithmetical* view of logic is based upon construing quantified propositions as enumerations of similar instances, so that "All Xs are Y" is thought of as an aggregate of singular propositions, "the first X is a Y, the second X is a Y, etc." The result is what De Morgan calls the *cumular* reading of quantified propositions (S, 166).

De Morgan's theory of the syllogism is strongly influenced by his method of generating all possible propositions. This is as true of the *Syllabus* as it is of his other writings on the syllogism, where it is based on three distinctions:

1. *Total vs. partial terms*: A term is used totally in a proposition when every instance must be examined to verify the proposition; it is used partially, when this need not be done. This distinction corresponds to the traditional contrast between distributed and undistributed terms.
2. *Affirmative vs. negative propositions*: A proposition is affirmative "which is true of X and X, false of X and not-X or x; negative, which is true of X and x, false of X and X." Thus, for example, "Every X is Y" is affirmative because "Every X is X" is true and "Every X is non-X" is false. Some propositions which look like denials are affirmative on this criterion. The proposition "Some things are neither Xs nor Ys" is affirmative because "Some things are neither Xs nor Xs" is true and "Some things are neither Xs nor non-Xs" is false.
3. *X vs. x*: The name "x" or "not-X," the contrary of X, applies to all objects within the universe of discourse to which the name "X" does not apply. It will be recalled that one reason for De

> Morgan's importance as a logician is his emphasis on the necessity of specifying the universe of discourse in interpreting any proposition, and requiring that for any X, both "X" and "x" name something.

The full spectrum of De Morgan's propositions can be generated from the first two distinctions, while the third provides alternative ways of saying the same thing. In terms reminiscent of his earlier account, De Morgan adopts the following "spicular" notation:

X) or (X	X is total.
X(or)X	X is partial.
.	sign of negative proposition.
no dot	sign of affirmative proposition.

There will then be eight propositional forms, depending upon whether X is total or partial, Y is total or partial, and the proposition is affirmative or negative (S, 158). These are:

Universal Propositions		*Contrary Particular Propositions*	
X))Y	All Xs are some Ys.	X(.(Y	Some Xs are not (all) Ys.
X).(Y	All Xs are not (all) Ys.	X()Y	Some Xs are some Ys.
X(.)Y	All things are either some Xs or some Ys (or both)	X)(Y	Some things are not either (all) Xs or (all) Ys.
X((Y	Some Xs are all Ys.	X).)Y	All Xs are not some Ys.

This categorization gives us the traditional A,E,I, and O propositions in X and Y, as well as in Y and X, expressed in a rather odd linguistic form. The only addition to the traditional scheme is the third line in each column; these are equivalent to "All x are Y" and "Some x are not Y," respectively.

These are all the propositions that can be constructed in this notation from the first two distinctions. If contraries are added to this system, one can insert into each of these eight forms one of four options XY, xY, Xy and xy, giving thirty- two propositional forms. Nevertheless, each of the new forms is equivalent to one of the original eight, so that no new forms are added (S, 158–59). A full set of logical relations between these forms of propositions can be determined, including contraversion, conversion, implication, and contradiction; but these need not be investigated at this point.

Having provided a theory of the proposition, and of immediate inference based on that theory, De Morgan moves to the syllogism. The same combinatorial strategy is applied to determine all possible forms of the syllogism in this notation. There are eight possible premises containing X and Y, and eight possible premises containing Y and Z; thus, there are sixty-four possible combinations of premises. Of these, thirty-two will turn out to have a valid conclusion (S, 160). De Morgan derives all these valid forms from the following four valid forms:

(6.1)	X))Y	All Xs are Ys.
	Y))Z	All Ys are Zs.
	X))Z	All Xs are Zs.
(6.2)	X()Y	Some Xs are Ys.
	Y))Z	All Ys are Zs.
	X()Z	Some Xs are Zs.
(6.3)	X((Y	Some Xs are all Ys.
	Y()Z	Some Ys are Zs .
	X()Z	Some Xs are Zs.
(6.4)	X((Y	Some Xs are all Ys.
	Y))Z	All Ys are Zs.
	X()Z	Some Xs are Zs.

Since negative propositions are equivalent to affirmative propositions about contraries, the full set of valid forms can be obtained by applying these forms to the eight possible combinations of terms and their contraries found in the two premises and the conclusion: that is, XYZ, xYZ, xyZ, xyz, xYz, XYz, Xyz and XyZ. Each of these forms can be put into an XYZ form by using the identities just outlined. Consider, for instance, the form,

(6.5)	x))y	All x are y.
	y))Z	All y are Z.
	x))Z.	All x are Z.

The identities translate this into,

(6.6)	X((Y	Some Xs are all Ys.
	Y(.)Z	Everything is either Y or Z.
	X(.)Z	Everything is either X or Z.

Six other valid forms can also be obtained from the first valid basic form, one for each combination of the three terms and their contraries. This can be done for all four basic forms, resulting in a total of thirty-two valid forms. An examination of cases shows that none of the other thirty-two combinations of premises can yield a valid conclusion. Instead of merely listing the valid and invalid forms, made systematic by De Morgan's spicular notation, one can provide sets of rules for judging the validity of syllogisms; but these rules need not be discussed here.

This method of obtaining and displaying the valid syllogistic forms has a certain elegance to it. For our purposes, the more important point is the means by which De Morgan demonstrates the validity of the four basic forms (S, 160). It is by this demonstration that he lays the groundwork for his relational construal of the arithmetical form of the cumular syllogism. Under the cumular interpretation, each quantified proposition asserts the "collection" of all its instances. Similarly, each syllogism asserts the collection of all the *unit syllogisms* which fall under them. If we consider the first of the four basic forms, and take each X in the universe, we can argue,

(6.7) This X is a Y.
That same Y is a Z.
―――――――――
This X is a Z.

Since this can be inferred for any X, "All X are Z" follows from the two original premises. This account is, by now, familiar. There is more than a hint of the logic of relations here, for each premise is really an identity assertion; and all that the conclusion requires is the transitivity of identity. The second basic form is based on the same unit syllogism, but with the understanding that it is being applied only to some X which is Y.

The third basic form requires both the transitivity and the convertibility of identity. Its unit syllogism will read,

(6.8) This Y is this X.
This Y is a Z.
―――――――――
This X is a Z.

or, to put it into proper form,

(6.9) This X is this Y.
This Y is a Z.
―――――――――
This X is a Z.

Convertibility of identity is needed for the first premise, after which transitivity can be applied. In the fourth form, the same unit syllogism is used, but with De Morgan's assumption of existential import for the term "Y."

One interesting feature of this arrangement is that De Morgan is concerned primarily with which premises give valid conclusions. Among those valid conclusions, he selects only the most powerful one. Thus, for example, he does not consider the conclusion "Some Xs are Zs" for the first basic form, even though the argument would be valid; but he does take this as the conclusion of the fourth form, since it is the strongest valid conclusion.

De Morgan outlines several complications to his system, such as the notation for "terminal precision" (so that, for instance, the notation "X)o)Y" will mean that both X))Y and Y))X), the distinction between cumular and exemplar readings of the standard propositions, and a system of the numerically definite syllogism. We may ignore these novelties, though, and consider his attempt to embed the arithmetical form of the cumular syllogism in his logic of relations.

3. OBJECTIVE VIEW–THE RELATIONAL FORM

De Morgan begins his discussion of the relational account by defining the four figures and noting that any syllogism can be stated in any of the four figures (S, 172). For the first basic form we have, for instance:

Figure I	*Figure II*	*Figure III*	*Figure IV*
X))Y	X))Y	Y((X	Y((X
Y))Z	Z((Y	Y))Z	Z((Y
X))Z	X))Z	X))Z	X))Z

De Morgan differs from most authors in writing his minor premise first. His intent in doing so is to suggest the composition of relations: X is related to Y, Y is related to Z, therefore X is related to Z. Furthermore, equivalent propositions are very easy to find in his system, so that equivalent syllogisms in other figures are readily available. In De Morgan's symbolism, a "converted" form of any proposition can be found by just reversing the order of the terms and making sure that each parenthesis continues to point in the right direction relative to each letter. Thus, in moving from Y))Z to Z((Y the order of the terms was changed, but the parentheses continued to point away from the Z and toward the Y. This can be done with all eight of De Morgan's propositional forms, and it is an elegant feature of his symbolism. It means that we can easily generate syllogisms in any "figure" we wish, even in cases where the English translation would make this unnatural. "Z((Y" puts the form under Figure II, even though its natural English translation, "All Y are Z," would have kept it in the first figure. We might, with De Morgan, read this as, "Some Zs are all Ys," thus keeping it in the second figure, but this is hardly colloquial English.

If every syllogism has a first figure form which can be easily determined, there will be little significance to figure in De Morgan's system. It only becomes important when De Morgan takes the "wider view" embodied in the doctrine of the abstract copula.

> The junction of names by appurtenance to one object, the copula hitherto used, is both convertible and transitive: and from these qualities, *and from these alone*, it derives the whole of its functional power in syllogism. Any copula which is both transitive and convertible will give precisely the syllogisms of our system, and no others: provided always that if contrary names be introduced, no instance of a name can, either directly or by transition, be joined by any copula with any instance of the contrary name. (S, 173)

De Morgan suggests that we take "is joined to" as the copula. Then the eight propositional forms become:

X))Y	Every X is joined to a Y.
X).(Y	No X is joined to a Y.
X(.)Y	Everything is joined to either an X or a Y.
X((Y	Every Y is joined to an X.
X(.(Y	Some X are not joined to a Y.
X()Y	Some X are joined to a Y.
X)(Y	Some things are joined neither to an X nor to a Y.
X).)Y	Some Y are not joined to an X.

This table is a straightforward translation of De Morgan's table of onymatic forms, including the third and seventh lines, which are analogues of De Morgan's two new propositional forms. There is an oddity which underlines the novelty of including these forms in the first place. It can best be seen by noting that if De Morgan is correct, the same propositional identities should hold for "joins" as for "is." This means that X))Y should be equivalent to x(.)Y. Yet, the former speaks of Xs being joined to Ys, while the latter speaks of things being joined to Xs or Ys. What is required in order to show that these forms are equivalent? We can utilize the three conditions which De Morgan stipulates:

1. "Is joined to" is transitive.
2. "Is joined to" is convertible.
3. No X is joined to an x, and no x is joined to an X.

This allows us to prove the implication in one direction. Assuming that x(.)Y, consider any a in X. Then it is joined to either an x or a Y. But by condition 3, it cannot be joined to an x, so that it must be joined to a Y. Since this holds for arbitrary a in X, X))Y.

When we move to the converse, something is lacking. The implication would not hold if there are isolated elements which are not joined to anything at all. A fourth condition is needed to preclude this possibility. This condition is not implausible, since it holds for the original case of identity.

4. Everything is joined to something.

The implication may then be demonstrated. Suppose that every X is joined to a Y, and that some c is not joined to any x. Then, since c is joined to something, it must be joined to some a in X. But this a must be joined to some b in Y. By the transitivity of "is joined to," the c is joined to some b in Y. Thus, any c must be joined either to x or to Y.

These conditions are also used in deriving the many other propositional equivalences involving "is joined to." For instance, let us derive X).(y from X))Y: that is, derive "No X is joined to a y" from "Every X is joined to a Y." Every a in X is joined to some b in Y; by condition 2, this b is joined to a. If a were then joined to some b′ in y, condition 1 would imply that the b in Y would be joined to some b′ in y. By condition 3, though, this is impossible. Thus, no a in X can be joined to any b′ in y; that is, no X is joined to a y.

The entire system of propositional equivalences for "is joined to" can be derived by this means. It can also be used to obtain the full theory of the syllogism. This requires only the four basic forms and the propositional equivalences. We have just seen how these equivalences can be demonstrated, and the four basic forms are clearly valid for "is joined to." Furthermore, the manipulation of these syllogisms into other figures can be accomplished as easily for "is joined to" as for "is identical with" (S, 174). Consider, for instance,

(6.10)	X))Y	Every X is joined to a Y.
	Y).(Z	No Y is joined to a Z.
	X).(Z	No X is joined to a Z.

If any a in X were joined to some c in Z, then c would be joined to a; but this a would also be joined to some b in Y. Thus, c would be joined to b, and b would be joined to a. But by the second premise, no Y is joined to a Z. Another of De Morgan's examples is,

(6.11)	X(.)Y	Everything is joined either to an X or to a Y.
	Y)(Z	Some things are joined neither to Ys nor to Zs.
	X(.(Z	Some Xs are not joined to Zs.

Every d is joined either to an a in X or a b in Y. Suppose it were joined to a, and that the conclusion is false–i.e., that every X is joined to a Z. Then a would be joined to some c in Z. By transitivity, this means that d would be joined to c. Thus, d would be joined either to some Z or to some Y. Since this would be true for any d, the second premise would be contradicted.

The special features of De Morgan's symbolism which make any first figure form transformable into a syllogism in any other figure–*in his notation*–apply equally well to "is joined to." Even when the translation is trivial, as in moving from X))Y to Y((X, it may depend on how the form is read. Both forms could be read as "Every X is tied to a Y," in which case there would be no inference at all. If we read it in line with De Morgan's first way of reading his onymatic notation, we would move from "Every X is joined to some Y" to "Some Y is joined to every X." The latter statement is ambiguous, though it is equivalent to the former statement on one reading. The convertibility of "is joined to" is required to infer from an a in X being joined to a b in Y that this b is joined to a.

A survey of possible cases supports De Morgan's claim that "The convertibility of the copula renders the inference altogether independent of figure," since within his notation, any inference can be transformed into an equivalent inference in any figure (S, 174). This holds for any system of quantified propositions which is obtained by reduction to unit propositions and syllogisms, with a copula which obeys the four conditions. The situation is more complex, though, when the convertibility condition is dropped (S, 174–75).

De Morgan's example of an inconvertible copula is "gives to," along with its converse "receives from." The propositional equivalences will now be expressed in a more complicated fashion, requiring both the relation and its converse, and their proofs will require more complex conditions on the relations. De Morgan lists these conditions as:

1. "Gives to" is transitive (so that "receives from" is also).
2. No X gives to another X, either directly or transitively, etc.
3. Every X either gives to a Y or receives from a y, but not both.
4. Every x either gives to a Y or receives from a y, but not both.
5. Every X which gives to a Y, receives from no other Y, etc.

In an annotation, De Morgan added a variant of the second condition,

6. No X gives to any X, either directly or transitively, etc.

The eight standard propositional forms must then be stated both in terms of "gives to" and its converse, with the added notation of "′" to express converse relations (S, 175).

X)′)Y Every X gives to a Y.
X(′.(Y Some Xs give to no Ys.
X)′.(Y No X gives to a Y.
X(′)Y Some Xs give to Ys.
X(′.)Y In every relation, something either gives to an X or receives from a Y (or both).
X)′(Y In some relations, nothing gives to any X nor receives from any Y.
X(′(Y Some Xs give to all the Ys.
X)′.)Y All Xs do not give to some Ys.

Similarly, there are eight propositional forms for "receives from," signaled by placing the "′" in a different place. These forms include, for instance,

X))′Y Every X receives from a Y.
X(.(′Y Some Xs receive from no Ys.
X(.)′Y In every relation, something either receives from an X or gives to a Y (or both).

The table of equivalent forms will use both relations and their converses. For instance, the following four forms are equivalent.

X)′)Y X).(′y x(.)′Y x(′(y

To illustrate, let us show that X)′)Y is equivalent to X).)′y; i.e., that every X gives to a Y just in case no X receives from a y. First, assume that X)′)Y. Condition 3 says that every X either gives to a Y or receives from a y, but not both. Since every X gives to a Y, no X will receive from a y. Conversely, condition 3 requires that if any X does not receive from a y, it must give to a Y. Thus, since no X receives from a y, every X must give to a Y.

This simple proof merely reflects the two aspects of condition 3. Noting that "gives to" is assumed to be transitive, we could have proved the first part of the equivalence by assuming that no y gives to a Y. For, suppose that a is in X, so that it gives to some b in Y. Suppose, also, that it received from some b′ in y. Then b′ would give to a. Since b′ gives to a and a gives to b, b′ would give to b; that is, some y would give to a Y, which would violate the condition that no y can give to a Y. Since this condition was required for the convertible copula, one wonders why it is not required for the inconvertible copula. It seems to be a more basic principle from which several of De Morgan's conditions follow.

This equivalence is simple, but others are more puzzling, and they raise the question of the meaning of De Morgan's "′" notation. Consider the equivalence between X)′)Y and x(.)′Y. The first means that every X gives to a Y, while, according to De Morgan, the second means, "In every relation, something either receives from an x or gives to a Y (or both)." The problem lies with the import of "in every relation." One could follow the lead of the ordinary copula, where this form would have been translated as "Everything is either an x or a Y." But is it the case that under the conditions specified, if every X gives to a Y, then everything either receives from an x or gives to a Y? It seems that this cannot be correct, since the Ys that receive from Xs will not necessarily receive from x or give to a Y. De Morgan, though, provides a proof:

> If every X give to a Y, the remaining Ys, if any, do not give to any Xs, by the assigned conditions of meaning: consequently those remaining Ys receive from xs [the contrary of X]. As to ys, none of them can give to Xs, for then they would give to Ys: therefore all receive from xs. Conversely, if x(.)'Y, no X can receive from y, for then neither could that y receive from x, nor could that X give to Y: so that there would be a relation in which neither does anything give to Y, nor receive from x. Consequently, every X gives to Y. (S, 175)

The first half of this proof raises the problem just stated. It shows that the "remaining" Ys, those which do not receive from an X, must receive from an x; and also, that the ys must receive from xs. But it does not show that the Ys which do receive from Xs must either receive from an x or give to a Y. We will leave this problem unresolved and move on to De Morgan's very brief treatment of syllogisms with inconvertible copulae (S, 176).

De Morgan provides his familiar list of the *phases* of syllogistic premises:

++ −+ +− −−

along with the distribution of phases in the four figures. The *Primary phases* for each figure are:

Figure	I	+	+
	II	−	+
	III	+	−
	IV	−	−.

He then states three rules, with the comment that "This addition is all that is required in the treatment of the syllogism of inconvertible copula." (S, 176) These rules are:

Rule 1. In the primary phases, the direct copula may be used throughout.

Rule 2. When one premise departs from the primary phase in quality, the converse copula must be used in the other.

Rule 3. When both premises depart from the primary phase in quality, the converse copula must be used in the conclusion.

Unfortunately, De Morgan provides only two examples of how this system is supposed to work, and its extension to other examples is unclear. But let us consider what is available. We begin with an example of the primary phase of figure II.[1] Such an argument would be:

(6.12)	X)′.(Y	No X gives to a Y.
	Z)′)Y	Every Z gives to a Y.
	X)′.(Z	No X gives to a Z.

The proof of this is simple. Suppose some X gave to a Z. Then by the second premise, this Z would give to a Y. By the transitivity of "gives to," the X would give to a Y, which is prohibited by the first premise. Let us now translate this into an equivalent form, but with different phases. In this situation, "equivalent" does not mean "logically equivalent," but rather, "equivalent under the five conditions governing the inconvertible copula." An equivalent form in which both premises differ from the primary phase in quality would be,

(6.13)	X)′)y	Every X gives to a y.
	Z)′.(y	No Z gives to a y.
	X).(′Z	No X receives from a Z.

Under the five conditions, each premise is equivalent to the corresponding premise of the primary phase argument, but of differing quality; by Rule 3, this means that the conclusion must contain the converse of the original relation, while remaining otherwise the same. The validity of this argument is easily seen. Suppose some X did receive from a Z. Then this Z would give to the X. By the first premise, the X would give to some y. From the transitivty of "gives to," Z would then give to some y. But this is precluded by the second premise.

The careful reader will note that one slip has been made in presenting this example. It is not true that, under the five conditions, each premise is equivalent to the corresponding original premise. The table of propositional identities does not state that X)′)y is equivalent to X)′.(Y; rather, X))′y is equivalent to X)′.(Y; and, similarly, for the second premise. De Morgan seems to have composed this paragraph in haste, for he not only leaves out the "′" sign to distinguish relations from their converses; he also changes the relation "gives to" to "gives." After presenting the first example, De Morgan writes merely of "contraverting" the middle term. Contraversion was introduced with the onymatic copula, for inferring from X))Y to X).(y; but its meaning with the generalized inconvertible copula is unclear. De Morgan's failure to add his accent here adds to the confusion.

The problem may be stated this way. De Morgan wants to take the primary phase as basic in each figure, and then use his rules to generate the other phases of that figure from the primary phase. The rules as stated, though, fail to take account of the fact that in moving from one relational proposition to another "equivalent" proposition, not only will quantity and distribution be changed; typically, the relation will be changed to its converse as well. In failing to change the relation to its converse, De Morgan has ignored an important complexity. In the absence of further instructions, we do not know how to generate the additional arguments. How, for instance, might one obtain an argument in which just one premise differs in quality from that in the primary phase? Consider the first premise of the original argument, X)′.(Y. The two forms which would change its quality would be X))′y and x((′Y. In both cases, the contrary of one term and the converse of the relation are taken. How, then, would the second premise, Z)′)Y, be adjusted to accommodate either of these options? If it is kept as it is, the second option would be used, in order to have the middle term the same in both premises. But then X becomes x. Will the conclusion contain X or x? On the other hand, if Z)′)Y is changed into its other positive form, z(′(y, the middle terms will match, but then it is unclear whether Z or z occurs in the conclusion.

From these and similar problems, I conclude that De Morgan's rules are not well-formulated and that they do not provide an adequate syllogistic system for the inconvertible copula.

4. THE SUBJECTIVE VIEW

Compared with the obscurities of the objective view, the subjective view provides the basis for an elegant relational treatment of the syllogistic system (S, 190–94).[2] The subjective view, it will be recalled, bases the syllogism on relations between either classes or attributes. De Morgan begins his account with his familiar views on relations. Logicians have traditionally evaded the full generality of the relational analysis of propositions by reducing all propositions to propositions of concomitance–that is, to the joint applicability of names. The result is the familiar reduction of all propositions and inferences to onymatic relations. Out of the onymatic relations, arise "five different modes of enunciating the same proposition." The first is the arithmetical mode, which we have considered in the objective view. The four subjective modes are (S, 190):

Mathematical: A class is contained in a class.
Physical: An attribute is predicated of a class.
Metaphysical: An attribute is a component of an attribute.
Contraphysical: A class is predicated of an attribute.

Thus for the proposition "humans are animals," we have:

Arithmetical: Every human is an animal.
Mathematical: The class human is contained in the class animal.
Physical: The class man has the attribute humanity.
Metaphysical: Animality is a component of humanity.
Contraphysical: All of which humanity is the attribute is in the class animal.

Every one of De Morgan's eight propositional forms can be interpreted in any one of these five ways. The crucial difference is this: on the *arithmetical* view, quantified propositions are taken as the conjunction of all their instances. Propositional equivalences are then proved by considering these individual cases; similarly, the validity of syllogisms is proved by reducing them to the unit syllogisms of which they are composed. On the other hand, in the *subjective* view, propositions, construed second intentionally, are already unit syllogisms, of which the terms stand for classes or attributes.

One way of underlining this change of standpoint involves the interpretation of De Morgan's spicular notation. In the *arithmetical* view, parentheses indicate whether or not a term is distributed in the proposition. The dot, or its absence, indicates the quality of the proposition. In the *subjective* view, the full spicular symbol, such as ").(", stands for a relation which holds between the two terms of the proposition. This means that each of the four subjective modes will allow eight possible relations. Aside from heuristic purposes, those relations need not even be indicated by the spicular notation; each relation could be expressed by a different primitive symbol.

De Morgan is saying, in effect, that the syllogistic calculus can be given at least five different interpretations. The four subjective interpretations are all of the same form; and the interpretation of the subject and predicate terms will determine the interpretation for the spicular relational terms. De Morgan does not work through all four interpretations, but considers only the mathematical and the metaphysical ones. We will focus our attention on the mathematical interpretation, since it is the most unproblematic. De Morgan provides names for each spicular relation, as he had in *FL*, but with changes of names (S, 191). These names are:

Symbolized Form	Arithmetical Objective Interpretation	Mathematical Subjective Interpretation	Symbol
X))Y	Every X is a Y.	X is a species of Y.	a
X((Y	Every Y is an X.	X is a genus of Y.	a*
X(.(Y	Some X are not Y.	X is an exient of Y.	o
X).)Y	Some Y are not X.	X is a deficient of Y.	o*
X).(Y	No X are Y.	X is an external of Y.	e
X()Y	Some X are Y.	X is a partient of Y.	i
X(.)Y	Everything is either X or Y.	X is a complement of Y.	u
X)(Y	Some things are neither X nor Y.	X is a co-inadequate of Y.	q

In this table, we have also added a symbol to represent each relation, for later use. It will be noticed that the last four relations are convertible, so that we may equally well say that X and Y are coexternals, copartients, complements and co-inadequates of each other.

Immediate inferences can then be interpreted in relational terms. One can say, for instance, that *genus* and *species* are converse relations, and that *species* and *exient* are contrary relations. The many propositional equivalences already noted can be read in this notation. For instance, the equivalence between X))Y and X).(y can be read as, "species is external of contrary," where "contrary" now refers to the contrary of a term. Most interesting, though, is the way in which this standpoint can be adapted to De Morgan's theory of the syllogism. The thirty-two valid forms of De Morgan's system can all be stated using the composition of relations. These valid forms are divided into two classes, of which twenty-four are regular syllogisms and eight are "strengthened" syllogisms (i.e., with premises stronger than needed for the conclusion). De Morgan divides the list of twenty-four into eight groups of three each, each triad consisting of propositions related by Theorem K. For instance, the first triad would be,

Species of species is species.
Genus of exient is exient.
Exient of genus is exient.

Using our symbols, these become instances of Theorem K:

aa))a	LM))N
a*o))o	L^{-1}n))m
o*a))o	nM^{-1}))l.

Rather than think of all of these relational propositions as inclusions, we can take seriously the notion of a relative product and treat these inferences like a multiplication table. Paul Lorenzen has taken this approach, but without taking into account the two additional propositional forms which De Morgan added to the traditional syllogism.[3] We can add these new forms and display the full table (see p. 168). To show the additions required by De Morgan's additional forms, the area of Lorenzen's more traditional table has been boxed off (though four new forms remain inside it). Noting that there would be no "u" or "q" within the traditional table, we obtain the rather surprising result that De Morgan has increased the number of valid forms from fifteen to thirty-two, merely by adding two new propositional forms.

The table embodies a rather different approach than does De Morgan's system of triads. De Morgan's triadic system takes three of the eight relations as primitive (e.g., species, external, and complement) and then defines the other relations in these terms. A few syllogistic forms are taken to be basic, and all the other forms are then derived using these definitions and Theorem K. In Lorenzen's formulation, on the other hand, each of the eight relations is taken as a primitive relation, and a multiplication table is then constructed for each pair of relations.

De Morgan's two new forms may seem to introduce something special into his system, but they do not introduce any genuinely new inferential forms. Their unusual verbal formulation is due to the fact that they are stated in terms of X and Y, just like all the other forms. But X)(Y, for instance, can equally well be read, "Some x are not Y." On this reading, the syllogism *eoq* becomes:

Table of De Morgan's valid forms

First Premise \ Second Premise	a	a*	o	o*	e	i	u	q
a	a	q	-	o*	e	-	o*	q
a*	i	a*	o	-	o	i	u	-
o	-	o	-	-	-	-	i	-
o*	o*	-	-	-	q	-	-	-
e	o*	e	q	-	q	o*	a	-
i	i	-	-	-	o	-	-	-
u	u	o	-	i	a*	-	i	o
q	q	-	-	-	-	-	o*	-

(6.14) No X are Y.
Some Y are not Z.
―――――――――――
Some X are not Z.

This is equivalent to,

(6.15) All Y are X.
All X are Z.
―――――――――――
All Y are Z. ,

which is just *aaa*.

This approach can be adapted to the other three subjective modes, of which De Morgan considers only the metaphysical mode (S, 194–96). The terms now denote attributes rather than classes, generating the following readings:

X))Y	X is a dependent of Y.
Y))X	X is an essential of Y.
X(.(Y	X is an independent of Y.
X).)Y	X is an inessential of Y.
X).(Y	X is a repugnant of Y.
X()Y	X is an irrepugnant of Y.
X(.)Y	X is an alternative of Y.
X)(Y	X is an inalternative of Y.

Once again, the last four relations are convertible. Immediate inferences and propositional identities can be generated. The first three syllogistic forms become,

Dependent of dependent is dependent.
Essential of independent is independent.
Independent of essential is independent. ,

being related, as before, by Theorem K.

In this chapter we have seen two ways in which De Morgan's theory of the syllogism can be viewed in the light of his logic of relations: in the arithmetical objective view, quantified syllogisms are reduced to unit syllogisms, which are established through the transitivity and the convertibility of the copula. In the subjective view, of whatever mode, these same syllogisms are seen to embody the composition of relations, without any reduction to unit syllogisms. The first approach corresponds to the doctrine of the abstract copula; the second, to the composition of relations.

We have also seen how De Morgan attempts to construct a theory of quantified syllogisms with inconvertible copulas which satisfy quite specific conditions. Since this also involves a reduction to unit syllogisms, it may be thought of as generalizing his first approach to the traditional syllogism. Unfortunately, the obscurities in this generalized account make comparisons difficult.

LOGIC AND MATHEMATICS

For over one hundred years, the links between logic and mathematics have been so close that it is difficult to think of the one without the other. All the persons chiefly responsible for the development of symbolic logic were both mathematicians and logicians.

Boole and De Morgan were important mathematicians of the Victorian era, and each made significant contributions to several fields of mathematics. The importance of mathematics for Boole's logical work is apparent from the very start, since he was primarily concerned with applying algebraic techniques to logic. The impact of mathematics upon De Morgan's logic is not so clear, however, and it is interesting to see how the two sides of his professional life were related.

These issues can be addressed from several different standpoints. We will begin with De Morgan's vision of a "mathematical logic" which unites logic and mathematics, and then we will consider the impact of three types of mathematical considerations upon De Morgan's logic, especially his logic of relations: the use of algebraic techniques and analogies, the analysis of geometrical proof, and the analysis of algebraic reasoning.

We will conclude our discussion with a sustained account of the mathematical roots of De Morgan's unusual conception of form as the "science of the action of the machinery." This conception of form is quite different from the logician's, and it stands for the results of a process of abstraction and generalization that De Morgan finds in mathematics. The extension of the laws of arithmetic to negative and imaginary ("impossible") numbers is the classic example of this process, and De Morgan's strong interests in the "formal" foundations of algebra had a profound impact on his conception of the logic of relations. For De Morgan, the doctrine of the abstract copula and, in fact, the whole logic of relations, represented just such a process of abstraction and generalization.

1. "A MATHEMATICAL LOGIC"

De Morgan became Professor of Mathematics at the new University of London (now University College) in 1828, only a year after graduating from Cambridge. With one five-year interruption, he stayed at University College until 1866. He is probably remembered best as a teacher of mathematics. His student W.S. Jevons was to say that as a teacher he was "unrivaled."[1] During this time, he taught several sequences of courses ranging from Euclid through differential equations. His teaching of mathematics was motivated by a strong belief in the importance of mathematics for training the mind of the general student, and his lectures showed a rigor which his students long remembered.

Many papers and textbooks flowed from De Morgan's pen, but it is fair to say that he was not a mathematician of great originality or power. Perhaps the sheer breadth of his intellectual interests, covering almost all aspects of Victorian culture, led him not to devote his full energies to mathematics.[2] In addition, the most profound work in mathematics was being done on the continent. Nevertheless, many of De Morgan's papers were clever, and his foundational papers showed a deep concern for putting mathematics on a rigorous basis. This is especially true of his best-known work, four papers entitled "On the Foundation of Algebra" (1839–44), in which he attempted to make sense of negative and complex numbers and to extend complex numbers to number triples.[3]

De Morgan never taught logic. He was Professor of Mathematics, and logic was not part of his responsibilities. His concern for logical rigor, and his belief that clear thinking is best learned through the study of mathematics, led him to stress logical issues in his teaching of geometry; but the paucity of logical materials in the vast number of teaching notebooks which he left behind, shows that this was but a small corner of his teaching.[4] Thus, De Morgan's extensive writings on logic did not emerge naturally from his normal professional responsibilities; rather, they constituted one part of the very extensive program of reading and writing which De Morgan conducted all his life outside of mathematics. It is ironic that De Morgan's most original work, and that which probably had the most lasting impact, was in a field that was something of an avocation for him.

De Morgan himself might not agree with this characterization, for he repeatedly emphasized the importance of uniting logic and mathematics. His work in these two fields may have seemed less separated to him than it does to us, when we compare De Morgan with such truly mathematical logicians as Boole and Frege. He was, in fact, something of a prophet of mathematical logic. It is likely that some of the energy behind his attacks on Sir William Hamilton came from their conflicting views of mathematics. Hamilton was the author of a well-known attack on the manner in which the Cambridge system of education emphasized the use of mathematics to teach intellectual rigor.[5] According to Hamilton, logic studies the necessary laws of thought, and mathematics applies these necessary laws to the study of a necessary matter–that is, the forms of space and time which are the foundations of geometry and algebra. The fact that geometry is a priori means that teaching mathematics is the worst possible way to teach logical skills, since those skills are to be used primarily with contingent matter.

De Morgan accepted Hamilton's premises, but drew the opposite conclusion.[6] In his inaugural address of 1828,[7] and in many subsequent writings, he claims that it is just the clarity of mathematical concepts, and the necessity of the propositions which contain them, that make mathematics the ideal means for inculcating rigor. In mathematics, the student can focus upon logical forms of inference, without being distracted by the ambiguities of our ordinary argumentation.

The controversy with Hamilton went deeper than mere pedagogy. De Morgan noted that almost everyone who had attempted to codify or extend logic had been a mathematician. In fact, he claimed that Hamilton had been the only non-mathematician to try this, and that his lack of mathematical training was one of the chief sources of the incoherence in his system (S5, 336). He was shocked that Hamilton's pronouncements on mathematics were accompanied by several egregious mathematical errors which showed that he was simply incompetent in mathematics.

In retrospect, De Morgan looked upon his controversy with Hamilton not as a personal controversy, but as a dispute over the place of mathematics in logic. De Morgan's interest was in uniting the two. In a footnote to *S3*, which followed his comparison of the ways in which logicians and mathematicians deal with the problem of form, he added:

I am fully aware of the boldness of my comparison of the logician and mathematician, and of the audacious appearance which it is likely to present to a class of inquirers who have hitherto been allowed to distribute functions to the branches of human knowledge pretty nearly in their own way. . . . As joint attention to logic and mathematics increases, a logic will grow up among the mathematicians, distinguished from the logic of the logicians by having the mathematical element properly subordinated to the rest. This *mathematical* logic . . . will commend itself to the educated world by showing an actual representation of their form of thought–a representation the truth of which they recognize–instead of a mutilated and onesided fragment, founded upon canons of which they neither feel the force nor see the utility. (S3,78)

Twelve years later he was to repeat the same theme.

I believe, and I am joined by many reflecting persons, among students of both logic and of mathematics, that as the increasing number of those who attend to both becomes larger and larger still, a serious discussion will arise upon the connexion of the two great branches of exact science, the study of the necessary laws of thought, the study of the necessary matter of thought. The severance which has been widening ever since physical philosophy discovered how to make mathematics her own special instrument will be examined, and the history of it will be written. (S5, 345)[9]

It should be emphasized that De Morgan did not think that this uniting of disciplines would benefit only logic. It would equally benefit mathematics, to which it would contribute heightened standards of rigor.

To the mathematician I assert that from the time when logical study was neglected by his class, the accuracy of mathematical reasoning declined. An inverse process seems likely to restore logic to its old place. The present school of mathematicians is far more rigorous in demonstration than that of the early part of the century; and it may be expected that this revival will be followed by a renewal of logical study, as the only sure preservative against a relapse. (S5, 337)[10]

De Morgan clearly believed that it was important to unite logic and mathematics. It is one thing, though, to advocate this in principle, and quite another to carry it out. This chapter will examine De Morgan's logic in some detail to determine just how close this union was and to see whether his logical writings were significantly affected by mathematical considerations. This will require that we discuss three quite separate issues:

a. The use of mathematical techniques and analogies in De Morgan's logical writings;
b. The influence of issues concerning mathematical proof upon his logic, especially the logic of relations; and,
c. His conception of the form-matter distinction, as it is found in both logic and mathematics.

In the course of this discussion, it will be useful to distinguish between geometry and algebra. De Morgan drew quite a sharp distinction between them, even though he supported the use of algebraic techniques in geometry and geometrical techniques in algebra.[11] In their pure forms, he considered them to be quite different types of mathematical disciplines. Whether this is completely true or not, we shall see that they bear upon his logical work in quite different ways.

2. ALGEBRAIC TECHNIQUES AND ANALOGIES IN LOGIC

De Morgan has often been thought of as one who introduced the mathematical method into logic; and, indeed, he thought of himself in that way. The most obvious sense in which this might be so would be through using recognizably mathematical techniques in logic. Strangely enough, De Morgan did not do so. He made no use of geometrical techniques at all, not even simple diagrammatic methods. But what of algebraic techniques? Having made significant contributions to the foundation of algebra, De Morgan might be expected to introduce algebraic techniques into logic, as Boole did. This is not the case.

The contrast between Boole and De Morgan is vivid. De Morgan admired Boole greatly. He had assisted Boole in the transition from schoolmaster to professor of mathematics, and he aided Boole's family in their attempts to secure a pension after Boole died. He and Boole had a warm correspondence for twenty-two years despite the fact that they never met.[12] He praised Boole in print on more than one occasion. In *S3*, for instance, he noted the charge that they had both introduced algebraic methods into logic; and, after calling Boole the greater culprit and himself the lesser one, he added in a footnote, "Not meant for extenuation; I wish I were the greater one" (S3, 82). In his encyclopaedia article "Logic" (1860), which discussed several attempts to generalize logic, he said that "Mr. Boole's generalization of the forms of logic is by far the boldest and most original of those of which we have to treat"–despite the fact that he also treated his own work. In both of these quotations, De Morgan seems to have anticipated the judgment of history that Boole was the greater logician (L, 255).[13]

Despite this admiration, De Morgan made no serious attempt to introduce Boolean techniques into his own work. The very substantial body of his published and unpublished writings contain no attempts to work through Boole's system in detail, or to extend it. This remains very puzzling. His reluctance to work along Boolean lines may have been due to his almost obsessive fear of being charged again with plagiarism. More likely, it reflected the fact that he had his own research program. In any case, he made it clear to Boole that, while he greatly admired Boole's work and saw some similarities, they were on quite different tracks.[14] When De Morgan received from Boole a copy of *The Mathematical Analysis of Logic* (1847), he replied graciously, but added:

> There are some remarkable similarities between us. Not that I have used the connexion of algebraical laws with those of thought, but that I have employed mechanical modes of making transitions, with a notation which represents our head work.[15]

He then carries out a fairly complex derivation from Boole in his own notation. He seems to have viewed his own techniques as more clearly approximating our natural modes of inference.

Although De Morgan did not utilize Boolean techniques, *FL* did contain two substantial contributions which might be construed as importing algebraic techniques into logic. One was the replacement of the theory of modality by *probability theory*, construed as a theory of degrees of belief. This and De Morgan's other works on probability theory were of some importance in the development of inductive logic, but they do not involve the use of algebraic techniques in deductive logic. The *numerically definite syllogism*, on the other hand, was an attempt to generalize the syllogism by adding numerical quantifiers (FL, 163–95). It is certainly a type of deductive logic, and it does use algebraic techniques. However, these techniques are much closer to generalized arithmetic than they are to modern algebra. As a logical method, the numerically definite syllogism had little influence, and De Morgan did not return to it in a serious way in his later work.

We may conclude, then, that De Morgan's deductive logic used algebraic techniques in only one instance, the numerically definite syllogism. At the same time, analogies between mathematical operations and logical operations may have played some role in his thinking. He often commented, for instance, on the analogy between eliminating a variable from two equations and eliminating a middle term from a syllogism (S2, 56). In addition, some of the operations which he performs upon relations are similar to those which may be performed upon mathematical ratios (S4, 228). The most striking parallels between mathematics and logic, though, are not cases of analogy at all; they are, rather, cases in which mathematical inferences are *instances* of his logical forms of inference. To see this, we must turn to considerations of mathematical proof.

3. LOGIC AND GEOMETRICAL PROOF

De Morgan was a strong advocate of mathematical rigor, both in conception and in deduction.[16] His interest in logic undoubtedly derived largely from his interest in mathematical reasoning. A natural hypothesis, then, is that major features of De Morgan's logic may have resulted from the detailed examination of mathematical proofs. We will test that hypothesis by first looking at geometry and then at algebra.

In Chapter One we juxtaposed two thousand years of reverence for Euclid and Aristotle and expressed amazement that in all that time no one had ever pursued seriously the question of whether Aristotle's logic sufficed for Euclid. We concluded that not even the first theorem of Euclid could be formally demonstrated prior to 1879.[17] Its demonstration requires the full resources of first-order quantification theory, with one-, two- and three-place predicates. Given the respect for Euclid as the paradigm of demonstrative science, it is ironic that symbolic logic developed largely from the attempt to formalize number theory and analysis, not geometry.

There is no doubt that De Morgan was seriously concerned with geometrical reasoning. One of the few attempts to cast a Euclidean proof in syllogistic form occurs in *SDM*. His first work in logic carried the title, "First Notions in Logic, Preparatory to the Study of Geometry." Furthermore, several short published works, along with the mathematical notebooks which he prepared for his classes, show a continuing interest in details of mathematical reasoning, and they contain a number of shrewd remarks.[18] He points out, for instance, that Euclid's lack of formal logic is shown on several occasions by his use of indirect demonstration. Having proved that *All S are P*, Euclid does not immediately infer *All non-P are non-S*. Rather, he assumes that something is non-P and S; and since it is S, it must also be P, which means that it is both P and non-P, which is a contradiction. This same move is made several times by Euclid, with no apparent realization that the general rule of contraposition is at work.

In addition to discussing details of geometrical reasoning, De Morgan suggests that Euclid helps to train our ability to discern formal principles of reasoning (S4, 240). The very fact that Euclid reasoned in such complex ways, but without the explicit apparatus of formal logic, means that students interested in discovering formal principles would be well-advised to look at Euclid from the logician's standpoint.

To what extent, though, did De Morgan himself do this? Not very much. "First Notions" was certainly *preparatory* to the study of geometry, since it contained no geometrical examples whatever; nor does *Formal Logic*. In fact, there exists in the entire corpus of De Morgan's work, published and unpublished, only one attempt at a detailed logical analysis of a geometrical proof. This is the analysis of a proof of the Pythagorean Theorem in *SDM* (221–23).

While De Morgan does not pursue the analysis of geometrical proofs, he does make some interesting general comments about the subject, the most significant of which are in *S4*. From this discussion it seems that he has a comparatively conservative view of the logical resources that are needed for geometry. He begins by noting that there appears to be a problem in applying logic to mathematics.

> It has sometimes been made a reproach to logic that the mathematicians, who reason more consecutively than any others when about their mathematics, do not regard the syllogism with respect in theory, and disdain it in practice. (S4, 239)

The relevance of syllogistic logic to mathematics does not require that every mathematical proof be stated in syllogistic form, any more than we need to think out the spelling of every word we use. It is important to know, though, that it can be done. "Those who cannot exhibit their inferences syllogistically need to learn; but those who can do not need to practice." (S4, 239) We may grant that geometrical proofs need not actually be stated in formal terms. The question is whether it is possible to do so. On this issue, De Morgan has quite definite views, though they are only stated and not justified.

> There is a useful but very limited field of exercise for the syllogism in geometry. There is hardly an instance, over and above the elimination of B from A = B, B = C, which is not an overt use of the *principium et exemplum*: whenever P is true, Q is true; in this case P is true; therefore in this case Q is true. The reduction into the pure technical form, except in a few instances at the commencement, would be useless. (S4, 240)

His emphasis here appears to be upon the usefulness of syllogistic reduction, not its possibility. He seems more concerned with pedagogy, especially when he later says that syllogistic reduction may be of "use to the learner," and that a few examples may even be usefully repeated "for practice" (S4, 240).[19]

As far as the adequacy of the syllogism is concerned, the tone of this passage is undoubtedly conservative; but De Morgan does note that other types of propositional transformations are needed as well. He makes the point, which we have already noted, that many of Euclid's indirect demonstrations are actually instances of the formal principle of contraposition. The syllogism must be supplemented by a variety of propositional transformations, of which contraposition is the chief example. In fact, the study of geometry "is of little, though some, account for technical exercise in the syllogism; of more for exercise in the transformations of the proposition" (S4, 241). He notes that the necessity of contraposition had long been known, so that its inclusion does not radically increase the stock of logical rules. These propositional transformations seem merely to be what is usually known as the theory of immediate inference.

All of this shows that De Morgan was closely tied to the tradition in two ways. First, the inclusion of some propositional transformations does not alter the fact that De Morgan adopted a basically syllogistic analysis of geometrical reasoning. It must be granted that he would construe the syllogism in relational terms; but, with one exception, the only part of relational logic which he thought geometry needed would be the syllogistic part. Secondly, he follows the tradition in failing to consider in detail the relationship between geometrical reasoning and formal logic. He seems to have accepted the view that a slightly augmented syllogism would suffice for geometry; but he did not seriously test this hypothesis.

It seems, then, that De Morgan's logical innovations were not seriously affected by issues concerning geometrical proof. However, there is one clause in his remarks which must give us pause. In the passage just quoted, he says, "There is hardly an instance, over and above the elimination of B from $A = B$, $B = C$, which is not an overt use of *principium et exemplum*." Geometrical reasoning can be reduced to the syllogism in all cases except those which involve the transitivity of equality. As we have seen in the first three chapters, this is a more serious qualification than appears at first sight.

It will be recalled that there had been one exception to the general lack of concern over the sufficiency of the syllogism for geometrical proof. This involved the problem of reducing to syllogistic form those inferences which require the transitivity of equality. We have seen that Thomas Reid denied that,

$$\text{(7.1)}\quad \frac{\begin{array}{c}B = C\\ A = B\end{array}}{A = C}$$

could be reduced to syllogistic form, but that Hamilton and Mansel said that it could be if a general material premise were added which asserted the transitivity of equality.

Problems concerning this inference may well have played a role in the development of De Morgan's doctrine of the abstract copula. Let us recall that in the chapter on geometrical reasoning in *SDM*, De Morgan adopts the conservative thesis that syllogistic logic suffices for geometrical reasoning. His account of the syllogism is quite traditional, except for one apparently minor point. He analyzes propositions into subject, copula and predicate; but he points out that there are two different copulas, *is* and *is equal to*. As a result, there are two quite different instances of 1 AAA: (7.1) and,

$$\begin{array}{l} \text{B is C} \\ \underline{\text{A is B}} \\ \text{A is C.} \end{array} \tag{7.2}$$

We have noted that there is no evidence that De Morgan, in 1831, was aware of the controversy surrounding this inference; and, he introduced it without commenting on its logical novelty. The best explanation for this is that even at this stage, De Morgan was giving a relational analysis of both copulas. While Reid would have drawn a radical distinction between the propositional forms in (7.1) and (7.2), De Morgan would say that they are both of the subject-copula-predicate form, since whatever is expressed by "is" is just as much a relation as equality.

To this approach, though, there is a natural objection: why stop with *is* and *is equal to*? Why are not the transitivity of "greater than" and "produces" equally good candidates? This is exactly the move which De Morgan made in *Formal Logic*, and the result is the abstract copula. What matters for traditional syllogistic is the fact that the traditional copula expresses a transitive and convertible relation; but any other transitive and convertible relation would do just as well. We merely postulate that an unnamed relation has the requisite logical properties and then deduce the moods and figures of the traditional syllogism from these purely formal considerations.

There are several stages in the development of De Morgan's logic of relations. We have seen how an issue in the logic of geometry may have led to the first step, the abstract copula. The view that *is* and *is equal to* are both copulas, and that they can both enter into 1 AAA inferences, seems to arise naturally from his analysis of geometrical reasoning. It is not plausible to limit this viewpoint to these two copulas alone, so that its natural generalization is the doctrine of the abstract copula. With some plausibility, then, this first step in De Morgan's logic of relations may well be related to problems of geometrical proof.

4. LOGIC AND ALGEBRAIC REASONING

What about the use of formal logic to represent algebraic inferences? Could any features of De Morgan's logic have arisen from his concern for algebraic reasoning?

These would seem odd questions to ask, in mid-nineteenth century terms. At the time, geometry was viewed as the paradigm of demonstrative science, being the clear deduction of theorems from evident and necessary first principles. Algebra, on the other hand, was not so much a science as a grab-bag of rules which stood in no clear relation to each other. Furthermore, the whole enterprise was of doubtful theoretical value because it trafficked in such dubious objects as negative and impossible quantities. Algebra had a disreputable air about it which made it odd even to ask about the logic of its reasoning. It was, in fact, computation and not reasoning at all.

De Morgan did not dispute this image of algebra, but he tried to change the reality behind it by putting algebra on a more rational foundation.[20] Since he consistently held that the main value of mathematics for the general student was to serve as an example of rigorous reasoning, the state of algebra could only be a scandal to him. He put algebra on a more rigorous footing in two ways: by determining the first principles of algebra and carefully deducing algebraic results from them; and by providing clear interpretations for those parts of algebra which raised conceptual problems.[21] In this program, of course, De Morgan was not alone, for he was working within the Cambridge tradition. The two parts of this program are related, for the extension of standard algebraic laws to negative and impossible quantities required a precise idea of just what those laws are.

We may call this project the systematization of algebra; we might even call it the axiomatization of algebra, if this did not have too many twentieth century connotations. Even here, though, there is little concern for the detailed logical analysis of reasoning. What we found with geometry applies equally to algebra: while De Morgan believed strongly in rigorous mathematics in the axiomatic tradition, he did not work out the logical problems in formally validating algebraic inferences. In fact, while De Morgan had worked through one geometrical proof in a reasonably formal way, he provides no algebraic examples at all.

De Morgan, then, did not attempt to analyze algebraic proofs in detail. Might it be the case, though, that general considerations concerning algebraic modes of inference had some impact on his thinking in logic? If the traditional syllogism dominated his thinking about geometrical proof, what about algebraic reasoning? We shall see that there is a sense in which the logic of algebraic reasoning is central to De Morgan's logical thinking, even though he does not provide any detailed analysis of it.

We have already noted that De Morgan saw certain analogies between logical and algebraic problems and operations. On one central point concerning the nature and fundamental laws of logic, though, he saw more than an analogy between the two disciplines; in fact, he presented algebra as the clearest *example* of his general theory of logic. To see this, we must return to the second stage of De Morgan's logic of relations, the composition of relations and the bicopular syllogism.

De Morgan generalized the abstract copula by allowing that different relations can occur in the two premises, resulting in inferences of the form:

(7.3)
yLz
xMy
———
xLMz
LM)) N
———
xNz

Both steps are universally valid, though the material premise that LM))N is needed for the second step. What had been dealt with through the abstract copula then becomes,

(7.4)
yLz
xLy
———
xLLz
LL))L
———
xLz,

which is just an instance of (7.3). In this case, the second inference requires the transitivity of L. The entire theory of the syllogism, both singular and quantified, can be embedded in this framework.

De Morgan claimed that the composition of relations is found throughout algebra. The composition of functions in algebra is more than analogous to the composition of relations: it is an instance of it. This is especially important because an entire branch of mathematics to which De Morgan made fundamental contributions, the Calculus of Functions, is based upon just such compositions.[22]

This relationship is made clear in De Morgan's first discussion of the composition of relations in *S2*. Having provided an example from ordinary language involving the composition of relations, De Morgan adds,

> The algebraical equation $y = \Phi x$ has the copula $=$, relatively to y and Φ x: but relatively to y and x the copula is $= \Phi$. This is precisely the distinction of 'John can persuade Thomas' and 'John is [one who can persuade Thomas.]' The deduction of $y = \Phi\Psi z$ from $y = \Phi x$ and $x = \Psi z$ is the formation of the composite copula $= \Phi\Psi$. And thus may be seen the analogy by which the instrumental part of inference may be described as *the elimination of a term* by *composition of relations*. For though in ordinary inference the concluding copula is usually identical with those premised, yet it is no less true that the composition must have taken place: X is Y, Y is Z, therefore X is that which is (= is) Z. (S2, 56)

While De Morgan speaks of analogy here, it is clear that something else is meant, for, with the formation of the composite copula, this inference is just of the form:

(7.5)	yLx	[L is $(= \Phi)$.]
	xMz	[M is $(= \Psi)$.]
	yLMz	[LM is $(= \Phi\Psi)$.]

In *S4*, the importance of algebra for this new approach to logic is made even more apparent. We have already seen how De Morgan paints a comparatively conservative picture of geometrical inference in *S4*. It is almost exclusively syllogistic, the only additions being traditional propositional transformations and the transitivity of equality. Algebra, though, is quite another matter.

> It is to *algebra* that we must look for the most habitual use of logical forms . . . so soon as the syllogism is considered under the aspect of combination of relations, it becomes clear that there is more of syllogism, and more of its variety, in algebra than in any other subject whatever, though the matter of the relations–pure quantity–is itself of small

variety. And here the general idea of relation emerges, and for the first time in the history of knowledge, the notions of relation and *relation of relation* are symbolized. . . . [T]he algebraist was living in the higher atmosphere of syllogism, the unceasing composition of relation, before it was admitted that such an atmosphere existed. (S4, 241)

A similar comment is to be found in *S3*, in which De Morgan replied to the charge that he had confused all of thought with algebra by saying,

He who makes me confound all other thought with algebra, because I call attention to what is more visible in algebra than in other thought, though it exists in all thought, must make his own logic responsible for the inference, not mine. (S3, 82)

We have now examined the relation between the first two stages in De Morgan's logic of relations and issues of mathematical proof. We have found a connection between the abstract copula and geometrical proof, and one between the composition of relations and algebraic inference. What, though, about the third stage, the full logic of relations, complete with the converse and contrary of relations, as well as inherent quantity? This is a very important issue, for in the first two stages De Morgan might be charged with putting old wine in new bottles, merely placing traditional logic in a new setting. In *S4*, though, he adds greatly to the stock of valid inferences to provide what might well be called a *logic* of relations. The logical principles involved are quite sophisticated, and one must admire De Morgan's ability to work them out in a fairly primitive notation.

From the standpoint of extending logic, rather than reformulating it, *S4* is undoubtedly De Morgan's most important work. Yet, there is almost no evidence of mathematical influences upon this novel and important memoir. The only algebraic analogy at work is the similarity between the converse of a relation and the inverse of a function. More importantly, there is no suggestion that the complex inferences which it formulates are required for mathematical reasoning. While one can assume that there are mathematical instances of these laws, no emphasis is placed upon them and no such examples are given.

5. FORM IN ALGEBRA AND LOGIC

The final connection between mathematical and logical ideas in De Morgan's work is philosophical in nature. This concerns the conception of form which plays such a fundamental role in the most novel and controversial elements of De Morgan's thinking about logic.

De Morgan maintained that both the mathematician and the logician are concerned with form, but that "the great distinction of form and matter is more in the theory of the logician than in his practice, more in the practice of the mathematician than in his theory" (S3, 78). He pointed out that there are levels of form in algebra, just as there are levels of form in the logical analysis of judgments. He claimed to be applying to logic a conception of form which was best seen by looking at the levels of generalization in algebra. To appreciate De Morgan's views on the nature of formal logic, it will be important to examine his views on the nature of algebra; and these, in turn, must be placed in the context of nineteenth century British algebra.[23]

Mathematics was traditionally conceived to be the "science of quantity." Developments in British algebra in the nineteenth century led to the very different view that it is the science which draws necessary conclusions from (perhaps) arbitrarily stated assumptions. This development of a "formalistic" philosophy of mathematics played an important role in De Morgan's mathematical and logical thinking, and the notion of form which it embodies is central to De Morgan's views on logic.

While this formalistic conception of algebra seems very natural today, it is important to remember that it did not arise in a vacuum. It arose as a response to the problems for algebra caused by the use of negative and imaginary (or "impossible") numbers. The use of these numbers presents a severe challenge to the view that mathematics is the science of quantity. Yet, their introduction is necessary in order to solve certain elementary equations. Negative numbers are needed to assure a solution to many first degree equations, such as $x + 5 = 3$, while imaginary numbers are needed for such second-degree equations as $x^2 + 1 = 0$. The general solution to second-degree equations requires that expressions of the form "$a + b\sqrt{-1}$" be meaningful.

It seems clear that there cannot be such a thing as a negative quantity. At the same time, the appearance of clarity about negative numbers can be provided by analogies such as the difference between having and owing money, or the contrast between a line pointing in the positive direction and one pointing in the negative direction. Imaginary numbers, though, are much more puzzling. What is an imaginary quantity? How could a real and an imaginary quantity be added? It seems that there is no such thing as a number which, when multiplied by itself, gives a negative number.

Nevertheless, imaginary roots appear to be genuine roots. They result from the use of standard formulas and, when substituted back in an equation, allow it to be verified. They are required to preserve the fundamental principle that every equation of the second degree has two roots. Perhaps the most puzzling feature of imaginary numbers is that the same laws govern them as govern real numbers. While we may have nagging doubts about their use, the fact remains that we can operate consistently with them in many contexts, using the same laws as we use for real numbers. When they are used at either the intermediate or final stages of solving equations, they introduce no inconsistency and the results can be directly verified. How can these strange symbols be understood?

The natural suggestion is that imaginary numbers be given a geometrical interpretation, since a geometrical interpretation makes sense of negative numbers. In 1806, geometrical interpretations were published by Búee and Argand, and these were later put into their now-standard form by Gauss. According to this interpretation, the complex number $a + bi$ (where "i" stands for $\sqrt{-1}$) stands for the directed line segment in the Cartesian plane whose x-coordinate is *a* and whose y-coordinate is *b*. Multiplying $(a + bi)$ by *i* has the effect of rotating the line counter-clockwise by ninety degrees; and multiplying again by *i* produces another ninety-degree rotation. Thus, the effect of multiplying twice by *i* is to produce a 180 degree rotation–that is, a congruent line pointing in the opposite direction. This fits neatly with the earlier interpretation of negative quantities, since i^2 is just -1, and negative quantities can be construed as an equal line pointing in the opposite direction.

The geometric interpretation of complex numbers helped to resolve many of the doubts about their legitimacy, but not everyone was convinced. William Frend, De Morgan's friend and eventual father-in-law, was never converted to either negative or imaginary numbers, and he wrote a book on algebra which carefully avoided them.[24] The geometrical interpretation did show that these "numbers" could be used consistently, but it is hard to see how their geometrical interpretation could resolve doubts about their meaning in algebraic contexts.

De Morgan always endorsed the use of negative and imaginary numbers on pragmatic grounds, though he was clearly concerned about their interpretation and consistency. In *SDM* and later works, he pointed out that negative or impossible roots for equations which represent word problems show that something is wrong with the problem. They mean, in effect, that the problem is poorly formulated because it has no solution. Thus, "$x^2 + 2x + 2 = 0$" represents the problem, What two numbers have a sum of minus two and a product of plus two? He says:

> In solving equations of the second degree, we have obtained a new species of result, which indicates that the problem cannot be solved at all. . . . It is therefore absurd to suppose that there is any quantity which x can represent, and which satisfies the equation $x^2 = -a^2$, since that would be supposing that x^2, a positive quantity, is equal to the negative quantity $-a^2$. The solution is then said to be impossible, and it will be easy to show an instance in which such a result is obtained, and also to show that it arises from the absurdity of the problem. (SDM, 149–50)

After presenting such an example, De Morgan concludes:

> We have shown the symbol $\sqrt{-a}$ to be void of meaning, or rather self- contradictory and absurd. Nevertheless, by means of such symbols, a part of algebra is established which is of great utility. It depends upon the fact, which must be verified by experience, that the common rules of algebra may be applied to these expressions without leading to any false results. . . . [A]t present let him [the student] proceed with the subject as a part of the mechanism of algebra, on the assurance that by careful attention to the rules laid down he can never be led to any incorrect result. (SDM, 151–52)

De Morgan concludes his discussion of negative and imaginary numbers by observing that there is a difference between negative and imaginary roots of equations: when a negative root is found, the problem may be reformulated (by changing the signs of the terms of the equation) so as to yield a positive root, but when the root is imaginary, this kind of repair is impossible.

De Morgan later adopted and modified the geometrical interpretation of complex numbers. Before considering this, though, let us consider the conception of algebra in which this was embedded. From this standpoint, the most important fact about imaginary numbers is that they obey many of the same laws that real numbers obey.[25] This suggests that there must be some way of separating the rules which govern mathematical expressions from the manner in which they should be interpreted. The result of pursuing this suggestion is the conception of algebra which is associated with nineteenth century British mathematics.[26]

It was under the influence of De Morgan's teacher, George Peacock, that this view of algebra became clearly formulated. In systematic views which culminated in the second edition of *Treatise on Algebra* (1842, 1845), but are best expressed in "Report on the Recent Progress and Present State of Certain Branches of Analysis" (1833), Peacock distinguished between what he called "Arithmetical Algebra" and "Symbolical Algebra."[27] Arithmetical algebra is a universal arithmetic, in which all signs are used in the same sense as in common arithmetic. Subtraction is limited to preclude negative and impossible quantities, and the roots of equations are severely limited. Peacock reviews the attempts of Frend and Baron Maseres to confine algebra to arithmetical algebra, and he rejects them because of the severity of the limitations which they entail. At the same time, he attacks those mathematical writers who attempt to minimize the gulf between arithmetical and symbolical algebra. Such writers make the fundamental error of thinking "that the operations of addition, subtraction, multiplication and division are used in one science and in the other in no sense which the mind may not comprehend by a practicable, though it may not be by a very simple, process of generalization."[28] For Peacock, there is no natural sense in which we can view the symbols and operations of symbolical algebra as a generalization of the principles of arithmetic; rather, we have obtained a new and strictly formal science. The operations of arithmetical algebra can be defined in the usual arithmetic terms, but this is not so for the operations of symbolical algebra. As a result, Peacock concludes,

> It is more natural and philosophical, therefore, to assume such principles as independent and *ultimate*, as far as the science itself is concerned, in whatever manner they may have been suggested, so that it may thus become essentially a science of symbols and their combinations, constructed upon its own rules, which may be applied to arithmetic and to all other sciences by interpretation: by this means, interpretation will *follow*, and not *precede*, the operations of algebra and their results; an order of succession which a very slight examination of their necessary changes of meaning, corresponding to the changes in the specific values and applications of the symbols involved, will very speedily make manifest.[29]

Thus, Peacock rejects both the attempt to limit algebra to arithmetical algebra and the attempt to found symbolical algebra on arithmetical algebra. Symbolical algebra is to be an independent science, studying the laws of combination of its own symbols which need not be interpreted arithmetically. Peacock does not completely sever the connection between arithmetical and symbolical algebra, though, because he maintains that arithmetical algebra necessarily suggests the principles of symbolical algebra. That is, symbolical algebra must be capable of being interpreted in an arithmetical way. Since arithmetical algebra suggests the principles of symbolical algebra, the symbolical algebra must satisfy what Peacock calls the Principle of the Permanence of Equivalent Forms:

Whatever form is algebraically equivalent to another when expressed in general symbols, must continue to be equivalent, whatever those symbols denote.[30]

Conversely,

Whatever equivalent form is discoverable in arithmetical algebra considered as the science of suggestion, when the symbols are general in their form, though specific in their value, will continue to be an equivalent form when the symbols are general in their nature as well as in their form.[31]

This view is summarized by saying,

In arithmetical algebra, the definitions of the operations determine the rules; in symbolical algebra, the rules determine the meaning of the operations, or more properly speaking, they furnish the means of interpreting them: but the rules of the former science are invariably the same as those of the latter, in as much as the rules of the latter are assumed with this view, and merely differ from the former in the universality of their application . . . [32]

This means that the truths of arithmetical algebra that do not depend upon any special properties of their values may be considered to be true in symbolical algebra. In light of this, these first steps toward a more formal conception of algebra retain close ties with arithemtic, with the proviso that the formal system which constitutes symbolical algebra must admit of a true arithmetical interpretation. This restriction was breached for the first time in 1843, when Sir William Rowan Hamilton discovered, or invented, quaternions. This three-dimensional analogue of complex numbers is possible only by not obeying the commutative law.

De Morgan's views on algebra are very much in this tradition. They are stated most clearly in Book II of his textbook *Trigonometry and Double Algebra* (1849), in which he begins by describing in general terms just what a "symbolic calculus" is. His object is to construct an algebra,

upon a basis which will enable us to give a meaning to every symbol *and combination of symbols* before it is used, and consequently to dispense, first, with all unintelligible combination, secondly, with all search after interpretation of combinations subsequently to their first appearance.[33]

His phrase "unintelligible combination" obviously refers to imaginary numbers, especially since Book I had carried a chapter on the "unexplained symbol $\sqrt{-1}$."

Any "science of symbols," to be adequately formulated, must present symbols that are properly distinguishable from each other, meanings that are clearly stated, and rules of operation that follow from the meanings assigned. We can imagine formal systems which are deficient in one or more of these respects, even a formal system with clearly stated symbols and operations, but with no assigned meanings. De Morgan then describes at length what such a calculus would look like:

> As soon as the idea of acquiring symbols and laws of combination, without given meaning, has become familiar, the student has the notion of what I will call a *symbolic calculus*; which, with certain symbols and certain laws of combination, is *symbolic algebra*; an art, not a science; and an apparently useless art, except as it may afterwards furnish the grammar of a science. The proficient in a symbolic calculus would naturally demand a supply of meaning. Suppose him left without the power of obtaining it from without: his teacher is dead, and he must *invent meanings* for himself. His problem is, Given symbols and laws of combination, required meanings for the symbols of which the right to make those combinations shall be a logical consequence. He tries, and succeeds; he invents a set of meanings which satisfy the conditions. Has he then supplied what his teacher would have given, if he had lived? In one particular, certainly: he has turned his *symbolic* calculus into a *significant* one. But it does not follow that he has done it in the way which his teacher would have taught him, had he lived. It is possible that many different sets of meanings may, when attached to the symbols, make the rules necessary consequences. (TDA, 92–93)

De Morgan then gives a simple example of multiple interpretations, and he shows how changes in the assignment of meanings can affect the significance of complex expressions, and how existing meanings may limit the significance of complex expressions.

De Morgan then traces the relations between *particular arithmetic* (using only numerals and no variables), *universal arithmetic* (Arithmetical Algebra), *Single Algebra* (using "impossible subtraction" or negative numbers) and *Double Algebra* (Symbolical Algebra). Universal arithmetic always requires that in the subtraction of b from a, b must not be greater than a. The question of interpretation is raised when this restriction is lifted in Single Algebra. A consideration of concrete magnitude solves the problem, for concrete magnitudes often require a "direction" as well as a number (TDA, 95). We cannot be asked to measure off ten feet without being told the direction, or to enter ten dollars on a company's books without being told whether it is income or expense. Under this interpretation, "+4" indicates four units in one direction, and "−4" indicates four units in the opposite direction. This can always be represented by points on a line with a zero-point. The phrase "Single Algebra" expresses the fact that this type of object may be represented in a one-dimensional space.

The importance of the transition from universal arithmetic to Single Algebra only becomes apparent when the symbolic calculus is used. This is even more true of the transition from Single Algebra to Double Algebra, in which imaginary quantities are allowed. De Morgan presents a symbolic algebra for Double Algebra, stressing that "*with one exception*, no word nor sign of arithmetic or algebra has one atom of meaning throughout this chapter, the object of which is *symbols, and their laws of combination*, giving a *symbolic algebra* which may hereafter become the grammar of a hundred distinct *significant algebras*"(TDA, 101). Only the sign of equality is initially given a meaning, which is that the symbols between which it is placed have the same resulting meaning. De Morgan then lists the signs and laws of his symbolic algebra, noting that "ordinary algebra contains all these symbols and all these rules, but its assigned meanings do not make *all* results significant"(TDA, 104). His task is to provide an interpretation which will make all results significant.

The most important feature of Double Algebra is that $(\sqrt{-1})(\sqrt{-1}) = -1$. That is, when $\sqrt{-1}$ is applied twice to any positive quantity, it has the effect of changing $+a$ into $-a$. There are many possible interpretations which satisfy this condition, but "common consent" places a restriction that is remarkably similar to Peacock's Principle of the Permanence of Equivalent Forms:

> our completely significant algebra shall be an *extension* of the defective system with which we commence: meaning, that so far as that system goes, significantly, it shall be a part of the new system. It would not help us, with reference to the mathematics now established, if fifty completely significant systems were produced, unless in one or more of them the same story were told as in the old algebra, so far as this last tells any story at all. We must have, if possible (and I am to show that it is possible), all that we do understand still understood in the same sense, with such enlargement of meaning as will give significance to symbols which we do not now understand. Accordingly, +1 and −1 are still to signify diametrically opposite units. (TDA, 109)

After showing that one-dimensional interpretations will not work, De Morgan points out that two dimensions may suffice, since "We can pass from a line to its opposite, not only *along the line*, but also by *supposing the line to turn round*"(TDA, 114). Thus, "$\sqrt{-1}$" denotes revolution through a right angle. The geometrical interpretation, when fully expounded, allows us to apply Double Algebra to non- geometrical cases. Where a complex number is the root of an equation, it is impossible to solve the equation in the terms stated. In other and more profitable cases, the answer is significant, but complex numbers have appeared in the process of solution. When this happens, the process of solution can be made intelligible by being translated into geometrical terms. Problems stated originally in terms of time, weight or other magnitudes can often be solved most easily by translating them into geometrical problems with the same form. The answer to the geometrical problem will then be the answer to the original problem, as long as the translation is faithful.

> The separation of essential from non-essential notions is a very important process to all who would think upon mathematical subjects. . . . The formation of symbolic algebra itself is a separation of the essential conditions of operation from the non-essential: the rejection of all meaning over and above the *points of meaning* on which transformations depend. (TDA, 114)

With these preliminaries out of the way, De Morgan provides a full geometrical interpretation for Double Algebra. Under this interpretation, the symbols which denoted numbers or magnitudes in Single Algebra are taken to denote lines, considered, in effect, as ordered pairs consisting of a length and a direction. The symbol "0" refers to an arbitrarily chosen zero-point, and "1" refers to a unit line. The sum of two lines, A + B, is the third side (length and direction) of the triangle formed by first drawing a line with the length and direction A, and then drawing one with the length and direction B from the end point of the first line. The product, A × B, is the line whose length is the product of the lengths of A and B, and whose angle from the horizontal axis is the sum of the angles of A and B. Each line is, then, an ordered pair consisting of a length and a direction (represented by an angle). Single Algebra is a part of this system, since it represents the restriction of Double Algebra to the x-axis. Standard theorems of Double Algebra can be given quite sophisticated geometrical interpretations which follow from the basic rules of interpretation (TDA, 121–22). The interpretation of exponents requires complications which we need not discuss (TDA, 129ff). It may easily be shown that every line may be represented in the form $A + B\sqrt{-1}$, which we now regard as standard. On this interpretation, the original laws of Double Algebra may all be verified and the principles of Trigonometry demonstrated. De Morgan concludes:

we have now a right to affirm that symbolic algebra is truly rendered significant by the preceding definitions; and that, so far, every identical equation of ordinary algebra is also an identical equation of double algebra. And further, that all ordinary or single algebra is so much of the double algebra as relates to the symbols of lines taken in the unit line or its continuation. (TDA, 125)

Let us return from this discussion of De Morgan's views on algebra to the question of their impact upon his philosophy of logic. What is meant by the talk of form in these contexts? We have already seen that two uses of the word "formal" apply to De Morgan's work. His logic includes the *formally valid* principles contained in the composition of relations and the logic of relations. Other principles, such as the transitivity of the copular relation and of "is tied to," are materially true principles, yet they deal with *formal properties* –i.e., properties which can be defined using only variables and logical words. In comparing logic and algebra, De Morgan is thinking of what is formal in yet a third sense, which we will call one of *abstraction and generalization*. It is this sense which De Morgan finds in the contrast between form and matter in both algebra and logic. If formal concerns are those which involve abstraction and generalization, then De Morgan's algebraic interests were formal, for they largely concerned the extent to which algebraic principles could be extended to the problematic cases of Single and Double Algebra. What justifies us in applying to negative and impossible quantities the same laws as we apply to regular quantities? To approach this issue one must first separate the form from the matter by distinguishing between formally specifiable laws and their several interpretations.

Instances of this concern with formality crop up elsewhere in De Morgan's work. What are we to make of such mysteries as "0/0," "infinity/infinity," and fractional exponents? Can some interpretation be given to the sum of a divergent series which would allow it to be handled algebraically?[34] De Morgan's interest throughout his algebraic work was in abstracting the laws which apply to clear cases, and then generalizing and interpreting them to include problematic cases.

De Morgan drew the parallel between algebra and logic explicitly, taking the former as the model for the latter. He explicitly compared the gradations of form and matter to be found in judgments and inferences with those to be found in algebra. In *S2*, he noted that algebra constitutes a generalization of arithmetic by saying,

In algebra, as it now stands, the *forms* born and educated in arithmetic have left their parent and set up for themselves. Any meanings which obey certain specified laws may be adopted as the means of giving expression to the forms: and the results must be accepted as true in every instance in which the combinations used are consistent and intelligible under the meanings given. (S2, 50)

Just as there can be many transitive copulas, so there can be many different types of objects represented by algebraic symbols. He goes on to draw the parallel explicitly.

In my work on *Formal Logic* . . . I followed the hint given by algebra, and separated the essential from the accidental characteristics of the copula, thereby shewing the conditions of invention for a copula different from the ordinary one, or for a copula which, being substituted for the ordinary one, shall leave all the forms and conditions of inference unaltered. (S2, 50)

The theme of the gradations of form and matter becomes evident in *S3*.

Distinctions which are of form in arithmetic become material in algebra. The lower forms of algebra become material in the algebra of the functional symbol. The functional form becomes material in the differential calculus, most visibly when this latter is merged in the calculus of operations. (S3, 77–78)

Finding the form, in logic as in algebra, means finding the formally specifiable principles of operation, of which our concrete inferences and calculations are instances.

The *form* or *law* of thought–asserted differences between these words being of no importance here–is detected when we watch the machine in operation without attending to the matter operated on. The form may again be separable into form of form and matter of form; and even the matter into form of matter and matter of matter; and so on. The *modus operandi* first detected may be one case of a limited or unlimited number, from all of which can be extracted one common and higher principle, by separation from details which are still differences of form. (S3, 75)

For De Morgan, then, finding form is watching "the machine in operation without attending to the matter operated upon." Nothing could describe better his conception of form in both logic and algebra.

6. CONCLUSIONS

Let us pull together the results of our examination of the relationships between the mathematical and the logical aspects of De Morgan's work. Our conclusions may be summarized as follows:

1. De Morgan's logic made little use of mathematical techniques and analogies.
2. De Morgan's logic owed little to the detailed analysis of mathematical proof or reasoning, in either geometry or algebra.
3. De Morgan's logic of relations has limited connections to his interests in mathematics; in particular,
   a. The doctrine of the abstract copula is a generalization of the transitivity of equality, which he saw to be required for geometrical proof;
   b. The composition of relations is most fully exhibited by algebra, especially by the calculus of functions;
   c. But, the full logic of relations, including inherent quantity, seems not to have been inspired by mathematical modes of thought.
4. Finally, the conception of form which lies behind the doctrine of the abstract copula and the composition of relations shows the process of abstraction and generalization which De Morgan took to be characteristic of algebra.

It is very difficult to compare the impact of these mathematical influences on De Morgan's thought with those deriving from logical and philosophical sources. But at least two other sources must be mentioned.

The first source is purely logical–De Morgan's interest in perfecting the formal theory of the syllogism. De Morgan was one of the great tinkerers in the history of logic, and he was always on the lookout for some new symbolism, some new and more elegant way of generating logical forms. One of the great attractions of the relational analysis of the syllogism must have been the opportunity it provided for elegant formulations–of both the traditional, or "onymatic" syllogism, and the more general theory of the bicopular syllogism.

The full logic of relations seems also to have been motivated by the desire to present the relational syllogism more elegantly. *S4* develops the logic of relations only to the point where it can be applied to the relational syllogism. De Morgan makes it clear that he was not able to complete his analysis of the relational syllogism until he developed a notation for inherent quantity.

A certain logical playfulness seems to lie behind *S3* and *S4* as well.

There is no suggestion that there is some circumscribed body of inferences which De Morgan is codifying, and for which he is claiming completeness. It is more that he is opening up a new field of inference to exploration, in the hope that others will enter this new land as well.

If many of the sources of De Morgan's logic are purely logical, more philosophical considerations played a central role as well. De Morgan's relational analysis of judgment, and the theory of inference to which it gave rise, must have been the backbone of his investigations into the relational syllogism and the full logic of relations. De Morgan's analysis implied that the logic of relations is not merely a part of logic; it is its core.

A RIGOROUS FORMULATION

In the fifth chapter, we examined the logic of relations which De Morgan published in 1860. This extremely important and suggestive work opened the door to the fully developed logic of relations. Nevertheless, it was untidy and incomplete. Its reader must wonder just what principles are being assumed and how strong the resulting system is.

In this chapter we will attempt to understand this logic better by formulating it rigorously and using it to prove all of De Morgan's results. The limits of this logic will then be probed by proving additional theorems within it and by supplementing it in various ways. The importance of adding the theory of identity will be shown, and a surrogate for quantification theory will be suggested. We will conclude this chapter by formalizing De Morgan's suggestion that a uniqueness operator be introduced.

1. BASIC ISSUES

The initial system, which will attempt to capture De Morgan's own logic, will be referred to as system D. The most serious problem in constructing D is the tension between historical and systematic considerations. We wish to formulate *De Morgan's* logic; yet, he does not present it systematically. In fact, the whole notion of a deductive *system* is foreign to De Morgan. This would require the explicit statement of a class of immediate inferences, together with rules for linking them to form derivations in the system. De Morgan provides none of this. His interest is in results rather than process. He has discovered a great many relational arguments and he wants to convince the reader of their validity. The principles which he uses in doing so may or may not be stated explicitly.

At the same time, it is possible to determine which principles De Morgan used, if only implicitly, and to systematize them. That will be our task in the first six sections of this chapter. In this reconstruction, we will attempt to mirror De Morgan's reasoning as closely as possible.

We will select our axioms and rules from the principles which he used, and our proof techniques will be very similar to his.

Our point of departure will be Tarski's classic article on the calculus of relations, in which he gives an elegant formulation of the system deriving from Peirce and Schröder.[1] Our task will be to investigate that part of this system which is due to De Morgan.

Tarski uses eleven special symbols, in addition to the usual symbols of standard logic. They are of two kinds: the absolute (or Boolean) symbols and the relative (or Peircean) symbols.

Type of Symbol	Boolean Symbols	Peircean Symbols
Relation constants	1 Universal relation	1′ Identity relation
	O Null relation	O′ Diversity relation
Unary operations	‾ Complementation	$^{-1}$ Converse
Binary operations	+ Addition	┼ Relative addition
	· Multiplication	; Relative multiplication
Binary relation	= Identity	-----

It is obvious that not all these symbols need be taken as primitive symbols; but Tarski is not interested in minimizing the number of primitive terms. The important point to note is that the Boolean and Peircean symbols are distinct, in that one set cannot be defined in terms of the other.

Tarski shows that the calculus of relations can be axiomatized in two ways. In the *first* way, these new symbols are added to standard logic (including quantification theory), along with axioms which can be regarded as definitions of the new symbols. Some axioms of this type would be,

$$(x)(y)[x\ R{+}S\ y \equiv (xRy \lor xSy]$$

$$(x)(y)[x\ R;S\ y \equiv (\exists z)(xRz \,\&\, zSy)]$$

$$(x)(y)[x\ R \dagger S\ y \equiv (z)(xRz \lor zSy)]$$

$$R = S \equiv (x)(y)(xRy \equiv xSy)\ .$$

Many of the theorems which follow from these axioms will contain no individual variables or quantifiers, but only relational variables. Examples are,

$$R = R \supset R{+}T = S{+}T$$

$$R+\bar{R}=1$$

$$R\cdot S=S\cdot R$$

$$R \dagger S=\overline{\bar{R};\bar{S}}\,.$$

This is probably the most natural way of constructing the calculus of relations, but it is not the only way. In fact, if our interest is primarily in the calculus of relations, it may seem circuitous to embed it within quantification theory. Perhaps it can be constructed directly. This is the *second* method for formalizing the calculus of relations, and it is the method to which Tarski devotes most of his classic article. In this method, no individual variables or quantifiers are used. All fifteen axioms are purely relational theorems of the system as developed by the first method. Formulas of this system are confined to relational identities and truth-functional combinations thereof. One of its complex axioms is, for instance,

$$(R;S).\,T=O\supset(S;T).\,R=O\,.$$

In formalizing De Morgan's logic of relations, the first question we must face is which of these two methods to adopt. We have already seen that De Morgan's own work suggests both methods. On the one hand, his standard expressions, such as "X..LY" and "X..LMY," contain both individual and relational variables. And, while they contain no explicit quantifiers, quantification theory is often implicit in their meanings and in derivations. Thus, it appears that De Morgan would be attracted to the first method. On the other hand, its use would require quantification theory, which goes far beyond De Morgan's own work. The second approach is, in fact, the one he most often takes. Having motivated certain principles by using the first method, De Morgan then uses purely relational inferences in a quite autonomous way. Most of his arguments can best be reconstructed by the second method, and this is the method we will adopt.

Let us begin on the level of vocabulary. Which of Tarski's symbols did De Morgan use? The only *Boolean* symbols which he uses are complementation and the binary relation of identity between relations. He does not use the universal and null relations; and, while he acknowledges the possibility of incorporating Boolean addition, he never pursues the matter. He uses many *Peircean* symbols, including some that Tarski does not consider. The converse and relative multiplication operations are central to his work; and, while he does not use relative addition, it could have easily been incorporated into his system. His two signs of inherent quantity are also Peircean, though Tarski does not use them. D does not include the identity and diversity relations. We will see later, though, that there are good reasons for expanding D to include them. Since De Morgan often deals with inclusion between relations, it will be important to have a symbol for inclusion as well as for identity.

These, then, are the symbols we will need, in De Morgan's notation:

$\bar{m}$	contrary (complement) of M
M^{-1}	converse of M
LM	L of an M
LM′	L of every M
L,M	L of only M
L)) M	L is included in M
L‖M	L is identical to M.

To what extent should we use De Morgan's own symbolism? Absolute purity would demand that we not deviate from it; yet, in some ways it is clumsy and not perspicuous. Complementation should be represented by an operation symbol, rather than by the use of a lower case letter, if for no other reason than the fact that complex expressions must be complemented. To put the language into a modern form, there should be an operation symbol for relative multiplication, and different types of symbols should be given to the two signs of inherent quantity and placed between the variables operated upon. Parentheses should be used for punctuation. We can retain "))", but "‖" can be replaced by the more familiar "=". The result is:

$\bar{M}$	contrary (complement) of M
M^{-1}	converse of M
L;M	L of an M
LeM	L of every M
LoM	L of only M
L)) M	L is included in M
L = M	L is indentical to M.

We will set up D in a standard and completely formal way. Fully explicit derivations will be given from time to time, either to provide examples of explicit proofs or to exhibit some interesting feature of D.

On many other occasions, theorems may just be stated or given abbreviated proofs.

We will deviate greatly from Tarski in our choice of axioms and methods of proof. Most of his theorems involve equations with the name of the null relation as one term. This is very different from De Morgan's approach. The null relation is never used. Many of his principles involve inclusion rather than identity, so that the axioms and rules that may be extracted from his practice are quite different from Tarski's. The result is really a *third* method for presenting the logic of relations, similar to that used recently by Tarski and Givant.[2]

One additional decision needs to be made before setting up D. Tarski embedded his original calculus of relations in a system of sentential logic. More recently, though, systems of relational logic (e.g., that of Tarski and Givant) have dispensed with sentential logic and allowed only relational inclusions and identities, without any sentential connectives at all. Which method should we adopt?

This issue has cropped up elsewhere in the history of logic. Lukasiewicz's important account of Aristotle's syllogistic formulated that theory within sentential logic. Smiley and Corcoran showed, though, that Aristotle's syllogistic could be formulated better without using sentential logic.[3] This had several advantages. The formulation was closer to the methods Aristotle actually used. It avoided assuming an implicit substructure of sentential logic. And, it was simpler.

A case could be made for saying that D should be formulated within sentential logic. De Morgan's recurrent use of "if..then.." in stating valid argument forms suggests a sentential account. Yet, he never uses complex sentential inferences and he seldom uses other sentential connectives. His logic of relations contains no mechanism for negation. It will be closer to De Morgan's own mode of argument if we avoid the sentential formulation; and it will certainly be simpler.

2. THE SYSTEM D

In formulating D, we will use "L," "M," "N" and "O" (with subscripts if needed) as metalinguistic variables ranging over relation terms; and the metalinguistic variables "F," "G" and "H" (with subscripts if needed) will range over sentences.

1. *Primitive Vocabulary*

   The primitive vocabulary consists of the following symbols:

   Relation symbols: R_1, R_2, ...

Unary operators: $^{-}$, $^{-1}$
Binary operator: ;
Binary relations:)) , =
Punctuations symbols: (,)

2. *Formation Rules*
   a. If L is a relation symbol, then L is a relation term.
   b. If L is a relation term, then $\bar{L}$ and L^{-1} are relation terms.
   c. If L and M are relation terms, then (L;M) is a relation term.
   d. If L and M are relation terms, then L)) M and L = M are sentences.

   We will delete parentheses when it is convenient to do so.

3. *Definitions*
   1. $LeM =_{df} \overline{(\bar{L};M)}$
   2. $LoM =_{df} \overline{(L;\bar{M})}$

4. *Axioms*

   To capture De Morgan's logic, we will find it necessary to use six axioms and eight rules, all of which De Morgan used.

   1. $\bar{\bar{L}} = L$
   2. $(L^{-1})^{-1} = L$
   3. $(L;M);N \,))\, L;(M;N)$
   4. $(\bar{L})^{-1} \,))\, \overline{L^{-1}}$
   5. $(L;M)^{-1} \,))\, M^{-1};L^{-1}$
   6. $(LeM);M^{-1} \,))\, L$

5. *Rules of Inference*
   1. From L)) M and M)) N infer L)) N.
   2. From L)) M infer $\bar{M} \,))\, \bar{L}$.
   3. From L)) M infer $L^{-1} \,))\, M^{-1}$.
   4r. From L)) M infer L;N)) M;N.
   4l. From L)) M infer N;L)) N;M.
   5e. From L = M infer L)) M.
   From L = M infer M)) L.
   5i. From L)) M and M)) L infer L = M.

6. *Derivability*

   The sentence H is *derivable* in D from the sentences $F_1, \ldots, F_n$ (the assumptions) if and only if there is a finite sequence of sentences $H_1, \ldots, H_m$, such that H_m is H and each H_i is either an axiom of D, one of the F_i, or follows from one or more of the preceding lines by a rule of inference. We will abbreviate this by "$F_1, \ldots, F_n \vdash H$." If H is derivable from the empty set of assumptions, then H is a *theorem* of D, which we abbreviate by "$\vdash H$."

3. PROPERTIES OF INCLUSION AND IDENTITY

De Morgan routinely proves theorems by replacing one relation term by another relation term which has been proved to be identical to it. Such an inference has the form: $L = M$, $F(L) \vdash F(M)$. Complete fidelity to De Morgan would suggest that we treat this rule as a primitive rule of inference. We will, however, treat it as a derived rule, since it can be established by using the rules for "=" and the properties of "))."

We may begin by showing that "=" is an equivalence relation.

DR3.1. $L = M, M = N \vdash L = N$

Proof:		
	1. L = M	Assumption
	2. M = N	Assumption
	3. L)) M	1, R5e
	4. M)) N	2, R5e
	5. L)) N	3, 4 R1
	6. M)) L	1, R5e
	7. N)) M	2, R5e
	8. N)) L	6, 7, R1
	9. L = N	5, 6, R5i

A similar proof easily yields,

DR3.2. $L = M \vdash M = L$.

These rules can then be combined with Axiom 1 (or 2) to give,

Th3.1. $\vdash L = L$

and thus,

Th3.2. $\vdash$ L)) L.

Only a few additional theorems will be needed to obtain the metatheorem at which we are aiming.

DR3.3. $L = M, L)) N \vdash M)) N$

Proof:		
1.	$L = M$	Assumption
2.	$L)) N$	Assumption
3.	$M)) L$	1, R5e
4.	$M)) N$	2, 3, R1

Similarly, we may prove,

DR3.4. $L = M, M)) N \vdash L)) N$

DR3.5. $L = M, N)) L \vdash N)) M$

DR3.6. $L = M, N)) M \vdash N)) L$

DR3.7. $L = M, N = O, L)) N \vdash M)) O$

DR3.8. $L = M, N = O, M)) O \vdash L)) N$.

We will also need to show that performing the same operations on identical relations yields identical results. Thus, we will have,

DR3.9. $L = M \vdash \bar{L} = \bar{M}$.

Proof:		
1.	$L = M$	Assumption
2.	$L)) M$	1, R5e
3.	$M)) L$	1, R5e
4.	$\bar{M})) \bar{L}$	2, R2
5.	$\bar{L})) \bar{M}$	3, R2
6.	$\bar{L} = \bar{M}$	4, 5, R5i

Similarly, we may prove,

DR3.10. $L = M \vdash L^{-1} = M^{-1}$

DR3.11. $L = M \vdash L; N = M; N$

DR3.12. $L = M \vdash N; L = N; M$.

The last two theorems can be combined to prove,

DR3.13. $L = M, N = O \vdash L; N = M; O$.

With these very basic results, we are now in a position to prove the metatheorems which are the goal of this section.

Metatheorem I. Let N_1 be a relation term containing the relation term L; and let N_2 be the relation term obtained from N_1 by replacing one or more occurrences of L in N_1 by M. Then,

$$L = M \vdash N_1 = N_2 \quad \text{and} \quad M = L \vdash N_1 = N_2.$$

Proof. The proof will be by strong induction on the number of operations in N_1, in addition to those in L. We will prove the result for the case in which $L = M$; the proof of $M = L$ is very similar.

α-case: $n = O$. Then $N_1 = L$ and $N_2 = M$, so that the desired result is just $L = M \vdash L = M$.

β-case: Suppose that the metatheorem is true for all relation terms with less than *n* operations. We will show that it is then true for all relation terms with *n* operations. There are three possible cases:

Case 1: $N_1 = \bar{O}_1$, where O_1 contains L; and $N_2 = \bar{O}_2$, where O_2 contains M. By the inductive hypothesis, we have, $L = M \vdash O_1 = O_2$. By DR3.9, $O_1 = O_2 \vdash \bar{O}_1 = \bar{O}_2$. This, in turn, implies that $L = M \vdash N_1 = N_2$.

Case 2: $N_1 = O_1^{-1}$, where O_1 contains L; and $N_2 = O_2^{-1}$, where O_2 contains M. The proof for this case is the same as that for Case 1, except that DR3.10 is used instead of DR3.9.

Case 3: $N_1 = O_1; O_2$, where either or both of O_1 and O_2 may contain L; and $N_2 = O_3; O_4$, with M being in one or both of O_3 and O_4. By hypothesis, $L = M \vdash O_1 = O_3$ and $L = M \vdash O_2 = O_4$. By DR3.13, $O_1 = O_3, O_2 = O_4 \vdash O_1;O_2 = O_3;O_4$. Thus, $L = M \vdash O_1;O_2 = O_3;O_4$.

This completes our proof. In practice, of course, we will not be much interested in theorems of the type given in MT I. We will want to replace one relation term by another identical relation term in sentences, and for this we will need the following result.

Metatheorem II. Let F be a sentence of D which contains one or more occurrences of L, and let G be the result of replacing one or more of these occurrences by M. Then, $L = M, F \vdash G$ and $M = L, F \vdash G$.

Proof: We will prove the first result; the second follows trivially from the first. F must be either $N_1)) N_2$ or $N_1 = N_2$, where L may occur in either or both of N_1 and N_2. We consider each case.

Case 1: F is $N_1)) N_2$. Then G is $O_1)) O_2$, and M may occur in either or both of O_1 and O_2. From MT I, $L = M \vdash N_1 = O_1$ and $L = M \vdash N_2 = O_2$. By DR3.7, $N_1 = O_1, N_2 = O_2, N_1)) N_2 \vdash O_1)) O_2$. Thus, $L = M, N_1)) N_2 \vdash O_1)) O_2$.

Case 2: F is $N_1 = N_2$ and G is $O_1 = O_2$. The proof that $L = M$, $N_1 = N_2 \vdash O_1 = O_2$ follows in a similar way, using DR3.1.

This metatheorem allows us to replace identical relation terms at will. It will be the primary proof technique in D, as it was for De Morgan. Its use will be referred to by "RR."

So far, we have used the rules of D as the foundation for the basic properties of "))" and "=," and for the Rule of Replacement. (We did use Axiom 1 (or 2) to show that $\vdash L = L$ and $\vdash L)) L$, but that was a mere convenience. We could just as easily have assumed "L)) L" as another axiom.) We will now use Axioms 1 through 5 to generate the full set of basic principles which De Morgan used.

We begin by demonstrating some useful variants on Rules 2 and 3.

DR3.14. $\bar{L})) M \vdash \bar{M})) L$

Proof: 1. $\bar{L})) M$ — Assumption
2. $\bar{M})) \bar{\bar{L}}$ — 1, R2
3. $\bar{\bar{L}} = L$ — A1
4. $\bar{M})) L$ — 2, 3, RR

Let us now adopt a convention for shortening our derivations. In this derivation, an instance of Axiom 1 was stated explicitly and used with RR. In derivations to come, we will not state the axiom that is used with RR unless clarity demands it. Thus, in proving DR3.14 we could have eliminated line 3 and justified the last line by writing "2, A1, RR." We will take the liberty of doing this with theorems as well as axioms. Similar derivations will yield,

DR3.15. $L)) \bar{M} \vdash M)) \bar{L}$

DR3.16. $L^{-1})) M \vdash L)) M^{-1}$

DR3.17. $L)) M^{-1} \vdash L^{-1})) M$.

It is now our task to strengthen Axioms 3 through 5 from inclusions to identities.

Th3.3. $\vdash \overline{L^{-1}})) (\bar{L})^{-1}$

Proof: 1. $(\overline{L^{-1}})^{-1})) \overline{(L^{-1})^{-1}}$ — A4
2. $(\overline{L^{-1}})^{-1})) \bar{L}$ — 1, A2, RR
3. $\overline{L^{-1}})) (\bar{L})^{-1}$ — 2, DR3.16

Step 2 embodies our convention for shortening derivations. This theorem combines with A4 to yield,

Th3.4. $\vdash (\bar{L})^{-1} = \overline{L^{-1}}$.

We can now easily obtain the identities for A3 and A5.

Th3.5. $\vdash M^{-1};L^{-1})) (L;M)^{-1}$

Proof;	1. $(M^{-1};L^{-1})^{-1})) (L^{-1})^{-1}; (M^{-1})^{-1}$	A5
	2. $(M^{-1};L^{-1})^{-1})) L;M$	1, A2, RR(twice)
	3. $M^{-1};L^{-1})) (L;M)^{-1}$	2, DR3.16

A5 and Th3.5 then suffice for,

Th3.6. $\vdash (L;M)^{-1} = M^{-1};L^{-1}$.

The converse of A3 is,

Th3.7. $\vdash L;(M;N))) (L;M);N$

Proof:	1. $(N^{-1};M^{-1});L^{-1})) N^{-1};(M^{-1};L^{-1})$	A3
	2. $((N^{-1};M^{-1});L^{-1})^{-1})) (N^{-1};(M^{-1};L^{-1}))^{-1}$	1, R3
	3. $(L^{-1})^{-1};(N^{-1};M^{-1})^{-1})) (M^{-1};L^{-1})^{-1};(N^{-1})^{-1}$	2, Th3.6, RR (twice)
	4. $L;(N^{-1};M^{-1})^{-1})) (M^{-1};L^{-1})^{-1};N$	3, A2, RR(twice)
	5. $L;((M^{-1})^{-1};)N^{-1})^{-1}))) ((L^{-1})^{-1};(M^{-1})^{-1};N$	4, Th3.6, RR (twice)
	6. $L;(M;N))) (L;M);N$	5, A2, RR(four times).

Along with A3, this yields,

Th3.8. $\vdash (L;M);N = L;(M;N)$.

4. DE MORGAN'S BASIC IDENTITIES

With these preliminaries out of the way, we can now demonstrate the relational identities which are among De Morgan's most important contributions to logic (S4, 222–24). We begin with *single-letter* identities.

Th4.1. $\vdash \bar{L} = (\overline{L^{-1}})^{-1}$

Proof:	1. $(\bar{L})^{-1} = \overline{L^{-1}}$	Th3.4
	2. $((\bar{L}^{-1})^{-1}) = (\overline{L^{-1}})^{-1}$	1, DR3.10
	3. $\bar{L} = (\overline{L^{-1}})^{-1}$	2, A2, RR

Similarly,

Th4.2. $\vdash L^{-1} = \overline{(\bar{L})^{-1}}$.

We come now to the *double-letter* identities which relate the three binary operations.

Th4.3. $\vdash$ LeM $= \bar{L}o\bar{M}$

Proof:	1. LeM $= \overline{\bar{\bar{L}};M}$	D1
	2. LeM $= \overline{\bar{\bar{L}}\bar{\bar{M}}}$	1, A1, RR
	3. LeM $= \bar{L}o\bar{M}$	2, D2, RR

This, of course, means that,

Th4.4. $\vdash$ LoM $= \bar{L}e\bar{M}$.

Next, we consider the identities that involve conversion.

Th4.5. $\vdash (LeM)^{-1} = M^{-1}oL^{-1}$

Proof:	1. LeM $= \overline{\bar{\bar{L}};M}$	D1
	2. $(LeM)^{-1} = (\bar{L};M)^{-1}$	1, DR3.10
	3. $(LeM)^{-1} = \overline{(\bar{L};M)^{-1}}$	2, Th3.4, RR
	4. $(LeM)^{-1} = \overline{M^{-1};(\bar{L})^{-1}}$	3, Th3.6, RR
	5. $(LeM)^{-1} = M^{-};\overline{\overline{L^{-1}}}$	4, Th3.4, RR
	6. $(LeM)^{-1} = M^{-1}oL^{-1}$	5, D2, RR

This theorem along with the application of Th4.3 to "$(\bar{M})^{-1}e(\bar{L})^{-1}$" quickly proves,

Th4.6. $\vdash (LeM)^{-1} = (\bar{M})^{-1}e(\bar{L})^{-1}$.

Similar proofs can be given for theorems involving the converses of o-statements.

Th4.7. $\vdash (LoM)^{-1} = M^{-1}eL^{-1}$

Th4.8. $\vdash (LoM)^{-1} = (\bar{M})^{-1}o(\bar{L})^{-1}$

The identities which involve the contraries of compound relations follow almost immediately from the definition of "e" and "o."

Th4.9. $\vdash \overline{L;M} = \bar{L}eM$

Th4.10. $\vdash \overline{L;M} = Lo\bar{M}$

$$\textit{Th4.11.}\ \vdash \overline{\mathrm{LeM}} = \bar{\mathrm{L}};\mathrm{M}$$

$$\textit{Th4.12.}\ \vdash \overline{\mathrm{LoM}} = \mathrm{L};\bar{\mathrm{M}}$$

Finally, we come to theorems which involve the contrary of the converse of a compound relation; in virtue of Th3.4, of course, this will be the same as the converse of the contrary of that relation.

$$\textit{Th4.13.}\ \vdash \overline{(\mathrm{L;M})^{-1}} = \mathrm{M}^{-1}\mathrm{o}(\bar{\mathrm{L}})^{-1} \qquad [= (\overline{\mathrm{L;M}})^{-1}]$$

Proof:

1.	$(\mathrm{L;M})^{-1} = (\mathrm{M}^{-1};\mathrm{L}^{-1})$	Th3.6
2.	$\overline{(\mathrm{L;M})^{-1}} = (\overline{\mathrm{M}^{-1};\mathrm{L}^{-1}})$	1, DR3.9
3.	$\overline{(\mathrm{L;M})^{-1}} = \mathrm{M}^{-1}\mathrm{o}\overline{\mathrm{L}^{-1}}$	2, Th4.10, RR
4.	$(\mathrm{L;M})^{-1} = \mathrm{M}^{-1}\mathrm{o}(\bar{\mathrm{L}})^{-1}$	3, Th3.4, RR

Similarly, we can show,

$$\textit{Th4.14.}\ \vdash \overline{(\mathrm{L;M})^{-1}} = (\bar{\mathrm{M}})^{-1}\mathrm{eL}^{-1} \qquad [= (\overline{\mathrm{L;M}})^{-1}]$$

$$\textit{Th4.15.}\ \vdash \overline{(\mathrm{LeM})^{-1}} = \mathrm{M}^{-1};(\bar{\mathrm{L}})^{-1} \qquad [= (\overline{\mathrm{LeM}})^{-1}]$$

$$\textit{Th4.16.}\ \vdash (\mathrm{LoM})^{-1} = (\bar{\mathrm{M}})^{-1};\mathrm{L}^{-1} \qquad [= (\overline{\mathrm{LoM}})^{-1}].$$

This completes the proof of all the identities which De Morgan presents. Later, we will see that other identities – even rather simple ones – follow from De Morgan's axioms. For the moment, though, we are limiting our attention to the logical laws which De Morgan actually stated.

5. THEOREM K

De Morgan's demonstration of Theorem K is one of the most interesting parts of his logic of relations, for it comes the closest to being a formal proof (S4, 224–26). Thus, it helps to reveal the logical laws which De Morgan thought would be needed for the logic of relations. We will give the proof in full.

$$\textit{DR5.1.}\ \mathrm{L;M}))\mathrm{N} \vdash \bar{\mathrm{N}};\mathrm{M}^{-1}))\bar{\mathrm{L}}$$

Proof:

1.	$\mathrm{L;M}))\mathrm{N}$	Assumption
2.	$\bar{\mathrm{N}}))(\overline{\mathrm{L;M}})$	1, R2
3.	$(\overline{\mathrm{L;M}}) = \bar{\mathrm{L}}\mathrm{eM}$	Th4.9
4.	$\bar{\mathrm{N}}))\bar{\mathrm{L}}\mathrm{eM}$	2, 3, RR

5. $\bar{N};M^{-1}\,))\,(\bar{L}eM);M^{-1}$	4, R4r
6. $(\bar{L}eM);M^{-1}\,))\,\bar{L}$	A6
7. $\bar{N};M^{-1}\,))\,\bar{L}$	5, 6, R1

This proof contains the first use of Axiom 6 in our reconstruction of De Morgan's reasoning. Steps 3 and 4 could have been eliminated if A6 had been stated in the equivalent form "$(\overline{L;M});M^{-1}\,))\,\bar{L}$" and adjustments made to Steps 5 and 6. But the use of "$\bar{L}eM$" is more intuitive. Let us convince ourselves that A6 is correct. What it says, for instance, is that if John loves every child of some parent of Mary, then John loves Mary. In standard logic, it becomes,

$$(x)(y)[(\exists z)((w)(wMz \supset xLw)\ \&\ yMz) \supset xLy]\,,$$

which is quantificationally valid.

The other half of Theorem K can be proved from a theorem which can be derived from Axiom 6.

Th5.1. $\vdash L^{-1};(LoM)\,))\,M$

Proof:	1. $(M^{-1}eL^{-1});(L^{-1})^{-1}\,))\,M^{-1}$	A6
	2. $(M^{-1}eL^{-1});L\,))\,M^{-1}$	1, A2, RR
	3. $((M^{-1}eL^{-1});L)^{-1}\,))\,(M^{-1})^{-1}$	2, R3
	4. $((M^{-1}eL^{-1});L)^{-1}\,))\,M$	3, A2, RR
	5. $L^{-1};(M^{-1}eL^{-1})^{-1}\,))\,M$	4, Th3.6, RR
	6. $L^{-1};((LoM)^{-1})^{-1}\,))\,M$	5, Th4.7, RR
	7. $L^{-1};(LoM)\,))\,M$	6, A2, RR.

With this theorem in hand, the proof of the other half of Theorem K is similar to that of DR5.1.

DR5.2. $L;M\,))\,N \vdash L^{-1};\bar{N}\,))\,\bar{M}$

Proof:	1. $L;M\,))\,N$	Assumption
	2. $\bar{N}\,))\,(\overline{L;M})$	1, R2
	3. $(\overline{L;M}) = Lo\bar{M}$	Th4.10
	4. $\bar{N}\,))\,Lo\bar{M}$	2, 3, RR
	5. $L^{-1};\bar{N}\,))\,L^{-1};(Lo\bar{M})$	4, R4l
	6. $L^{-1};(Lo\bar{M})\,))\,\bar{M}$	Th5.1
	7. $L^{-1};\bar{N}\,))\,\bar{M}$	5, 6, R1

So far, our formal discussion of De Morgan's logic of relations has been quite unproblematic: the principles we have found in this logic have been clearly valid, and our task has only been one of bringing order to a somewhat disordered series of laws. At this point in De Morgan's presentation, though, the situation changes. We have seen in Chapter 5 that in his discussion of what follows from the *identity* of L;M and N, he asserts a problematic principle and fails to recognize a simpler and uncontroversial one.

De Morgan begins by noting that, "Identity, in Theorem K, does not give identity" (S4, 225). He means by this that the following analogues of DR5.1 and 5.2 do not hold:

$$\mathrm{L;M = N} \vdash \bar{\mathrm{N}}\mathrm{;M}^{-1} = \bar{\mathrm{L}}$$
$$\mathrm{L;M = N} \vdash \mathrm{L}^{-1}\mathrm{;}\bar{\mathrm{N}} = \bar{\mathrm{M}}.$$

De Morgan suggests that we look at the demonstrations to see why these principles do not hold. All of the steps in the proofs for analogues of DR5.1 and DR5.2 would go through for "=" in place of "))," except Step 6. In the proof of DR5.1 this is Axiom 6, while in the proof of DR5.2 it is Th5.1, which is deduced from Axiom 6. Why, then, is Axiom 6 only an inclusion and not an identity? An identity would require an inclusion in the opposite direction, which would be,

$$\mathrm{L\,))\,(LeM);M}^{-1}.$$

This can be translated into standard logic as,

$$(x)(y)[xLy \supset (\exists z)((w)(wMz \supset xLw)\ \&\ yMz)].$$

This expresses a strong assumption which should not be a law of logic. It means that if John loves Mary, then Mary has a child and John loves every parent of that child. This principle is doubly wrong: first, because Mary's having a child does not follow from John's loving Mary; and secondly, even if there were a child, it would not follow that John loves every parent of that child.

Having rightly rejected this line of argument, De Morgan proceeds to allow a problematic existential assumption by which he can deduce $\mathrm{L\,))\,N;M}^{-1}$ and $\mathrm{M\,))\,L}^{-1}\mathrm{;N}$ from L;M = N. In fact, only L;M)) N is needed for this result, not L;M = N. We shall indicate the dubious status of these laws by writing "?" before their number.

?DR5.3. $L;M\,))\,N \vdash L\,))\,N;M^{-1}$

Proof:		
1.	$\mathrm{L;M\,))\,N}$	Assumption
2.	$\mathrm{(L;M);M^{-1}\,))\,N;M^{-1}}$	1, R4r

3. $L;(M;M^{-1}))) N;M^{-1}$	2, Th3.8, RR
4. $L)) L;(M;M^{-1})$	New Principle
5. $L)) N;M^{-1}$	3, 4, R1

From this, of course, the principle which De Morgan actually enunciates follows trivially.

?DR5.4. $L;M = N \vdash L)) N;M^{-1}$

A very similar theorem can be deduced if we allow ourselves the principle that $L)) (M^{-1};M);L$, which can be derived from the new principle used in proving ?DR5.3.

?DR5.5. $L;M)) N \vdash M)) L^{-1};N$

?DR5.6. $L;M = N \vdash M)) L^{-1};N$

But what are we to make of this new principle, that $L)) L;(M;M^{-1})$? Its quantificational translation is,

$$(x)(y)[xLy \supset (\exists z)(\exists w)(xLz \;\&\; (zMw \;\&\; yMw))].$$

This, in turn, means that if John loves Mary, then there is someone whom John loves who is a parent of someone of whom Mary is a parent. The only way to guarantee this would be to make Mary be the parent whom John loves; and once again, of course, it does not follow from the fact that John loves Mary that Mary is a parent.

This new principle is only plausible when we assume that the *y* to which *x* stands in the relation L must stand in the relation $M;M^{-1}$ to itself–for instance, that Mary stands in the relation (Parent; Child) to herself. This, in fact, is the form in which De Morgan states the principle when he says that $(M;M^{-1})X$ is a class which must contain X. But, to say that $X(M;M^{-1})X$ always holds is to say that,

$$(x)(\exists y)(xMy \;\&\; yM^{-1}x),$$

that is,

$$(x)(\exists y)xMy.$$

This is a strong assumption which does not hold in general. Mary need not be a parent of a child, for she need not be a parent at all. Assuming this condition means restricting our attention to relations and universes of discourse which are such that every member of that universe stands in every relation to some member on other. Such a condition is obviously too restrictive, and one can only wonder why De Morgan held to it. Perhaps he was led astray by the fact that this condition does hold for some relations. For instance, while not every person stands in the relation (Parent; Child) to herself or himself, each does bear the relation (Child; Parent) to himself or herself. The notation for the inverse operation, suggesting the notation for functional inverses (e.g., multiplication and division), might also have been a factor.

We conclude this section by noting that De Morgan could easily have proved two theorems of the form "L)) . . " from the two parts of Theorem K. From DR5.1 and D1 we obtain,

DR5.7. L; M)) N ⊢ L)) NeM^{-1};

and, from DR5.2 and D2,

DR5.8. L;M)) N ⊢ M)) L^{-1}oN.

Each of these could equally well have "(L;M) = N" as its assumption. It seems odd that if De Morgan were looking for rules of the form "L;M = N ⊢ L)) . . .", he did not use these two rules rather than utilize "rules" which require such a strong existential assumption.

6. PROPERTIES OF RELATIONS

De Morgan discusses two of the three properties which are commonly attributed to relations, transitivity and convertibility (i.e., symmetricality) (S4, 226–27). He does not discuss reflexivity in the same formal way, though it is sometimes implicit in what he says. We will see later that the formal theory of reflexive relations can be developed within De Morgan's logic if we adopt a symbol for the identity relation.

We begin, then, with De Morgan's laws for *transitive* relations. Transitivity is easily defined in De Morgan's notation,

Df6.1. L is transitive = df L;L)) L.

We will abbreviate this by "T(L)," which will serve as the assumption for all our theorems about transitive relations.

DR6.1. T(L) ⊢ (L;L);L)) L

Proof:	1. L;L)) L	Assumption
	2. (L;L);L)) L;L	1, R4r
	3. (L;L);L)) L	1, 2, R1

De Morgan states a remarkable number of inclusions utilizing "L^{-1}" and "$\bar{L}$." We begin with the transitivity of L^{-1}, while noting that the transitivity of L does not imply the transitivity of $\bar{L}$.

DR6.2. $T(L) \vdash L^{-1};L^{-1})) L^{-1}$

Proof:		
1.	$L;L)) L$	Assumption
2.	$(L;L)^{-1})) L^{-1}$	1, R3
3.	$L^{-1};L^{-1})) L^{-1}$	2, Th3.6, RR

We can then obtain some more complex theorems, using Theorem K.

DR6.3. $T(L) \vdash L^{-1};\bar{L})) \bar{L}$

Proof:		
1.	$L;L)) L$	Assumption
2.	$L^{-1};\bar{L})) \bar{L}$	1, DR5.2

Using these results, we can also show,

DR6.4. $T(L) \vdash \bar{L};L^{-1})) \bar{L}$

DR6.5. $T(L) \vdash L;(\bar{L})^{-1})) (\bar{L})^{-1}$

DR6.6. $T(L) \vdash (\bar{L})^{-1};L)) (\bar{L})^{-1}$.

So far we have dealt only with theorems in which a single "L," suitably operated upon, appears to the right of the inclusion sign. De Morgan sees that there are many inclusions with a single "L" on the left side as well.

DR6.7. $T(L) \vdash L)) LeL^{-1}$

Proof:		
1.	$T(L)$	Assumption
2.	$\bar{L};L^{-1})) \bar{L}$	1, DR6.4
3.	$L)) \overline{\bar{L};L^{-1}}$	2, DR3.15
4.	$L)) LeL^{-1}$	4, D1, RR

From DR6.4 we can also get,

DR6.8. $T(L) \vdash L)) \bar{L}o(\bar{L})^{-1}$,

while DR6.3 yields,

DR6.9. $T(L) \vdash L)) (\bar{L})^{-1}e\bar{L}$

DR6.10. $T(L) \vdash L)) L^{-1}oL$.

These results can be used to generate inclusions with "L^{-1}" and "$\bar{L}$" on the left side of the inclusion sign.

DR6.11. $T(L) \vdash L^{-1})) LoL^{-1}$

Proof:		
	1. $T(L)$	Assumption
	2. $L)) LeL^{-1}$	1, DR6.7
	3. $L^{-1})) (LeL^{-1})^{-1}$	2, R3
	4. $L^{-1})) (L^{-1})^{-1}oL^{-1}$	3, Th4.5, RR
	5. $L^{-1})) LoL^{-1}$	4, A2, RR

Using DR 6.8–10, we can prove, in turn,

DR6.12. $T(L) \vdash L^{-1})) \bar{L}e(\bar{L})^{-1}$

DR6.13. $T(L) \vdash L^{-1})) (\bar{L})^{-1}o\bar{L}$

DR6.14. $T(L) \vdash L^{-1})) L^{-1}eL$.

Finally, there are theorems utilizing "$\bar{L}$" on the left side of the inclusion sign.

DR6.15. $T(L) \vdash \bar{L})) \bar{L}eL$

Proof:		
	1. $L;L)) L$	Assumption
	2. $\bar{L})) \overline{(L;L)}$	1, R2
	3. $\bar{L})) \bar{L}eL$	2, Th4.9, RR

De Morgan seems not to have noticed that the following theorem also holds,

DR6.16. $T(L) \vdash L)) Le\bar{L}$.

Equally easily, we can derive,

DR6.17. $T(L) \vdash L)) Lo\bar{L}$

and its companion,

DR6.18. $T(L) \vdash L)) \bar{L}oL$.

We can complete De Morgan's laws for transitive relations by proving,

DR6.19. $T(L) \vdash (\bar{L})^{-1})) L^{-1}o(\bar{L})^{-1}$.

Proof:		
	1. $T(L)$	Assumption
	2. $\bar{L})) \bar{L}eL$	1, DR6.15
	3. $(\bar{L})^{-1})) (\bar{L}eL)^{-1}$	2, R3
	4. $(\bar{L})^{-1})) L^{-1}o(\bar{L})^{-1}$	3, Th4.5, RR

Similarly,

DR6.20. $T(L) \vdash (\bar{L})^{-1})) (\bar{L})^{-1}eL^{-1}$.

This concludes our discussion of the laws governing transitive relations.

De Morgan's discussion of *convertible* (symmetrical) relations yields more interesting speculations than it does theorems. We define,

Df6.2. L is convertible = df $L \,)) \, L^{-1}$,

which we abbreviate as "C(L)." The inclusion is easily made an identity.

DR6.21. $C(L) \vdash L = L^{-1}$

Proof:		
1.	$L \,)) \, L^{-1}$	Assumption
2.	$L^{-1} \,)) \, (L^{-1})^{-1}$	1, R3
3.	$L^{-1} \,)) \, L$	2, A2, RR
4.	$L = L^{-1}$	1, 3, R5i

It follows from this that $L;L^{-1}$ is always convertible, no matter what L is.

Th6.1. $L;L^{-1} = (L;L^{-1})^{-1}$

Its proof needs only the use of Axioms 5 and 2. But, neither LeL^{-1} nor LoL^{-1} is convertible. Instead, from Th4.5,

Th6.2. $(LeL^{-1})^{-1} = LoL^{-1}$.

The inclusions which we have proved for transitive relations suggest that additional results could be obtained for relations that are both transitive and convertible, for the distinction between L and L^{-1} can be neglected for convertible relations. Oddly enough, De Morgan does not take up this point. But let us do so anyway.

The most perspicuous way in which to combine transitivity and convertibility is to ask the following question: How many logically distinct terms can be obtained by using some relation, L, twice, along with an arbitrary number of uses of "$^{-}$" and "$^{-1}$", together with one use of the relative product? To answer this question, we note first that each term can occur in one of five ways: L, $\bar{L}$, L^{-1}, $(\bar{L})^{-1}$, and $(\overline{L^{-1}})$. Furthermore, such a term could be operated upon in any number of these ways. At this point, the number of combinations threatens to get out of hand, but Axioms 1 and 2 and Theorem 3.4 step in to save the day. By Th3.4 any even number of uses of complementation can be brought into immediate proximity, and then eliminated by Axiom 1; similarly, excess inverses can be eliminated with Axiom 2. Furthermore, Th3.4 means that the last two basic forms are equivalent, leaving us with only four basic forms.

Each of the two terms in the relative product can thus have one of these four forms; furthermore, the entire relative product can also have any one of the four forms. This means that there will be 4^3, or 64, logically distinct forms that can be constructed in this way.

This number is drastically reduced to eight when we confine ourselves to *convertible* relations. There are now only two distinct terms, L and $\bar{L}$, with the four possible combinations L;L, $\bar{L}$;L, L;$\bar{L}$, and $\bar{L};\bar{L}$. To reduce the operations on these four forms, we can note that when L is convertible, we are not led outside these four forms by using conversion.

DR6.22. $C(L) \vdash (L;L)^{-1} = L;L$

Proof:		
	1. $C(L)$	Assumption
	2. $L = L^{-1}$	1, DR6.21
	3. $(L;L)^{-1} = L^{-1};L^{-1}$	Th 3.6
	4. $(L;L)^{-1} = L;L$	2, 3, RR(twice)

We can also show,

DR6.23. $C(L) \vdash (L;\bar{L})^{-1} = \bar{L};L$

DR6.24. $C(L) \vdash (\bar{L};L)^{-1} = L;\bar{L}$

DR6.25. $C(L) \vdash (\bar{L};\bar{L})^{-1} = \bar{L};\bar{L}$.

No such identities exist when we take the complements of relative products, so that when L is convertible we are left with eight distinct forms:

$$\begin{array}{llll} L;L & \bar{L};L & L;\bar{L} & \bar{L};\bar{L} \\ \overline{L;L} & \overline{\bar{L};L} & \overline{L;\bar{L}} & \overline{\bar{L};\bar{L}}. \end{array}$$

Two other laws for convertible relations, which have been implicit in what we have done so far, can be stated explicitly.

DR6.26. $C(L) \vdash C(\bar{L})$

DR6.27. $C(L) \vdash C(L^{-1})$

No new reductions of these forms can be obtained when we add *transitivity* to convertibility. Nevertheless, striking simplifications occur in DR6.2–6.20, since there is no longer a distinction between L and L^{-1}. It is surprising that De Morgan never noted these laws, since he insisted on reducing onymatic syllogisms to unit syllogisms with a transitive and convertible copula. It would seem only natural that he would consider the result of conjoining these two conditions. From DR6.3–6.6, we can easily obtain,

DR6.28. $\mathrm{C(L), T(L)} \vdash \mathrm{L;\bar{L}}\ ))\ \mathrm{\bar{L}}$

DR6.29. $\mathrm{C(L), T(L)} \vdash \mathrm{\bar{L};L}\ ))\ \mathrm{L}$.

These inclusions cannot be turned into identities, so that no further reduction in the number of distinct forms is possible. It will be possible to do so, though, when reflexivity is added to transitivity and convertibility.

It is interesting to note that there is no similar inclusion with "$\mathrm{\bar{L};\bar{L}}$" on the left side of the inclusion. This is so because the transtivity of L does not imply the transitivity of $\mathrm{\bar{L}}$. Even more obviously, $\mathrm{\bar{L};\bar{L}}$ need not be included in L, when L is transitive.

The equivalence of L and L^{-1} allows a great many permutations of the theorems we have proved, including those which involve "e" and "o"; but there would be little point in writing them all out. We should note that from the definitions of transitivity and convertibility, it trivially follows that,

DR6.30. $\mathrm{C(L), T(L)} \vdash \mathrm{L;L^{-1}}\ ))\ \mathrm{L}$.

And, from DR6.28, we can obtain,

DR6.31. $\mathrm{C(L), T(L)} \vdash \mathrm{L;(\overline{L^{-1}})}\ ))\ \mathrm{\bar{L}}$.

Both of these are commonly cited properties of equivalence relations.

7. ADDITIONAL INCLUSION LAWS

Our exposition of De Morgan's logic of relations is at an end. We have proved in D every law of that logic which he asserts. At the same time, we cannot help but wonder whether the system can be extended beyond the point to which De Morgan developed it. There are three possibilities for such an extension.

1. We can prove theorems of D which De Morgan did not assert.

2. We can find principles of the logic of relations which do not follow from D, but which can be added to it to obtain an enlarged system.
3. We can add new notations to D, along with new axioms and/or rules governing those notations.

We have already seen that De Morgan did not record some fairly obvious theorems of D. There are undoubtedly a large number of other unrecorded theorems of D which are important for understanding its structure. We will confine our attention to a few theorems which tidy up the inclusion relation.

Rules 4r and 4l can be combined to yield the more general principle,

DR7.1. L))M, N))O ⊢ L;N))M;O.

Proof:	1. L))M	Assumption
	2. N))O	Assumption
	3. L;N))M;N	1, R4r
	4. M;N))M;O	2, R4l
	5. L;N))M;O	3,4, R1

In fact, DR7.1 could have been taken as a rule from which Rules 4r and 4l could be derived. While this procedure might have been more elegant than assuming two separate axioms, it would have reflected De Morgan's own reasoning less well.

Equivalent forms of this theorem may be stated in terms of the two signs of inherent quantity.

DR7.2. L))M, N))O ⊢ LeO))MeN

Versions of Rules 4r and 4l, using "e," then follow easily.

DR7.3. L))M ⊢ LeN))MeN

DR7.4. L))M ⊢ NeN))NeL

Similar theorems for "o" may be proved as well.

DR7.5. L))M, N))O ⊢ MoN)) LoO

DR7.6. L))M ⊢ MoN))LoN

DR7.7. L))M ⊢ NoL))NoM

8. THE FULL SYSTEM OF THREE-RELATION TERMS

In section 8.6, we adopted one type of natural stance toward D, the combinatorial standpoint. In that section, we asked how many logically distinct relations could be constructed from one relation by applying the operations of D in a limited way; and we also saw how imposing transitivity and convertibility conditions dramatically reduced the number of distinct forms. From the indefinitely large number of other combinatorial problems, we will select the following one: What are all the logical relations between terms that can be constructed using three distinct letters and two occurrences of binary operation symbols, without complementation or conversion?

Let us begin with the letters L, M and N and the two operation symbols ";" and "e." Aside from variations that arise from permuting the order in which the relations are written, there are eight possible forms which can be constructed in this way. Their logical relations are given in the following table:

$$L;(M;N) = (L;M);N$$

$$(LeM)eN = Le(M;N)$$

$$L;(MeN)\,)) \,(L;M)eN$$

$$(LeM);N\,))\, Le(MeN).$$

The first identity is simply Th3.8. The definition of "e" allows us to deduce the second identity from the first.

Th8.1. $\vdash$ $(LeM)eN = Le(M;N)$

The first inclusion on this list follows from A6.[4]

Th8.2. $\vdash$ $L;(MeN)\,))\,(L;M)eN$

Proof:	1. $(MeN);N^{-1}\,))\,M$	A6
	2. $L;((MeN);N^{-1})\,))\,L;M$	1, R41
	3. $(L;(MeN))\,;N^{-1}))L;M$	2, Th3.8, RR
	4. $L;(MeN)\,))\,(L;M)e(N^{-1})^{-1}$	3, DR5.7
	5. $L;(MeN)\,))\,(L;M)eN$	4, A2, RR

The other inclusion follows from this theorem.

Th8.3. $\vdash$ $(LeM);N\,))\,Le(MeN)$

Proof:	1. $\bar{L};(MeN)\ ))\ (\bar{L};M)eN$	Th8.2
	2. $\overline{(\bar{L};M)eN}\ ))\ \overline{\bar{L};(MeN)}$	1, R2
	3. $(\overline{\bar{L};M});N\ ))\ \overline{\bar{L};(MeN)}$	2, Th4.11 ,RR
	4. $(LeM);N\ ))\ \overline{\bar{L};(MeN)}$	3, D1, RR
	5. $(LeM);N\ ))\ Le(MeN)$	4, D1, RR

Similar results can be deduced for all the three-relation terms using ";" and "o." Theorem 3.8 and the definition of "o" can be used to prove,

Th8.4. ⊢ Lo(MoN) = (L;M)oN.

Then Th8.2 is used to prove the two inclusions,

Th8.5. ⊢ (LoM);N)) Lo(M;N)

Th8.6. ⊢ L;(MoN))) (LoM)oN.

Finally, we can consider the relationships between three-relation terms which use only "e" and "o." We have already considered all but four cases:

Lo(MeN) = (LoM)eN

Le(MoN)

(LeM)oN.

Theorem 3.8, together with the definitions of "e" and "o" suffice to prove the identity,

Th8.7. ⊢ Lo(MeN) = (LoM)eN.

However, no equivalence or inclusion holds for the other two terms within the class of terms which we have been considering.

If we had expanded the combinatorial problem to include the unary operations on relations, many more relationships would have arisen. They hardly need to be catalogued, but two are of interest because they involve the "leftovers" from our last case.

Th8.8. ⊢ $Le(MoN) = \bar{L}o(M;\bar{N})$

Th8.9. ⊢ $(LeM)oN = (\bar{L};M)e\bar{N}$

Theorem 8.2 is of considerable interest in its own right, for some interesting theorems can be derived from it. For instance, there is,

DR8.1. L;M)) N ⊢ L;(MeO))) NeO.

Proof:	1. L;M)) N	Assumption
	2. (L;M)eO)) NeO	1, DR7.3
	3. L;(MeO))) (L;M)eO	Th8.2
	4. L;(MeO))) NeO	2, 3, R1

Th8.10. ⊢ (LeM);(M^{-1}eN))) LeN

Proof:	1. (LeM);M^{-1})) L	A6
	2. (LeM);(M^{-1}eN))) LeN	1, DR8.1

This theorem can be construed as a generalization of Axiom 6. It is quite a powerful device, for it allows us to "cancel out" terms. It can be expanded to any number of clauses, so that successive applications of it will give, for instance,

Th8.11. ⊢ ((LeM);(M^{-1}eN));(N^{-1}eO))) LeO.

The same phenomenon occurs with "o," for which there is a theorem with the same form as Th8.10.

Th8.12. ⊢ (LoM);(M^{-1}oN))) LoN.

This concludes our discussion of the System D. We will now see what happens when we allow ourselves the luxury of making some additions to it.

9. DE MORGAN'S LOGIC WITH IDENTITY

In this section we will add a symbol for the identity relation to D in the hope of getting some further insights into the potentialities of De Morgan's logic of relations. This addition would seem natural to De Morgan. He repeatedly emphasized the centrality of the identity relation in traditional logic, and he claimed that the traditional syllogism rests merely on the formal properties of that relation. His recognition of the identity relation as one important relation among many relations suggests that he would welcome its addition to his logic.

From a modern standpoint, we recall that it is a standard practice to add the theory of identity to quantification theory. Since D is a system of two-place predicate logic with some implicit quantification, it is of interest to see what D with identity would look like. Furthermore, the logics of relations which have derived ultimately from De Morgan (e.g., those of Peirce, Schröder, and Tarski) all contain the identity relation.[5]

The use of the identity relation allows us to formulate a variety of conditions which can be formulated in standard quantification theory but not in D alone. The result is a considerable expansion of G in the direction of becoming a full theory of quantification. The definition of reflexivity can be expressed, and a surrogate for quantification theory can be developed. The apparatus of D can then be applied to this enlarged system.

The system of De Morgan's logic with identity, DI, can be obtained from D by adding "I" to the vocabulary and making one obvious addition to the formation rules. Only one axiom is needed:

Axiom I. $\mathrm{L;I = L}$.

The fact that I is an *equivalence relation* follows easily from Axiom I. Its *transitivity* arises from substituting "I" for "L" in AI.

*Th*9.1. $\vdash \mathrm{I;I = I}$,

from which transitivity follows,

*Th*9.2. $\vdash \mathrm{I;I}\,))\,\mathrm{I}$.

Convertibility, which we state in the form of an equivalence, requires a bit more work.

*Th*9.3. $\vdash \mathrm{I}^{-1} = \mathrm{I}$

Proof:	1. $\mathrm{I}^{-1};\mathrm{I} = \mathrm{I}^{-1}$	AI
	2. $(\mathrm{I}^{-1};\mathrm{I})^{-1} = (\mathrm{I}^{-1})^{-1}$	1, DR3.10
	3. $\mathrm{I}^{-1};(\mathrm{I}^{-1})^{-1} = (\mathrm{I}^{-1})^{-1}$	2, Th3.6, RR
	4. $\mathrm{I}^{-1};\mathrm{I} = \mathrm{I}$	3, A2, RR(twice)
	5. $\mathrm{I}^{-1} = \mathrm{I}$	4, 1, RR.

The third property of an equivalence relation is that it be *reflexive*. In section 8.6, we noted that transitivity and convertibility could be easily formulated in D, but that there was no apparent way to formulate reflexivity. With the addition of "I," this can be done easily. To say that a relation is reflexive in a universe of discourse is to say that every element of that universe stands in that relation to itself. Since every element of any universe stands in the identity relation to itself, the identity relation is included in every reflexive relation. We can thus define,

Df9.1. L is reflexive = df I)) L.

We will abbreviate this by "R(L)." From Th3.2 it follows trivially that I is reflexive.

Th9.4. $\vdash$ I))I

I has other interesting properties. AI easily yields,

Th9.5. $\vdash$ I;L = L.

I also functions as a partial identity on "e" and "o."

Th9.6. $\vdash$ LeI = L

Proof:		
	1. LeI = $\overline{\bar{L};I}$	D1
	2. $\bar{L}$;I = $\bar{L}$	AI
	3. LeI = $\bar{\bar{L}}$	1, 2, RR
	4. LeI = L	3, A1, RR

Similarly,

Th9.7. $\vdash$ IoL = L.

It should be noted that I is only a partial identity for "e" and "o." L is not identical to either IeL or LoI.

To show how the definition of reflexivity can be used, let us prove,

DR9.1. R(L) $\vdash$ M))L;M

Proof:		
	1. I))L	Assumption
	2. I;M))L;M	1, R4r
	3. M))L;M	2, Th9.5, RR

We can also prove,

DR9.2. R(L) $\vdash$ M))M;L.

Intuitively, these principles makes sense. If L is reflexive, then, if x bears M to y, there is something–namely, x itself–which is such that x bears L to it and it bears M to y.

10. MORE PROPERTIES OF RELATIONS

In section 8.6, we concluded that there are eight logically distinct forms of compound relations for a convertible L:

$$L;L \qquad \bar{L};L \qquad L;\bar{L} \qquad \bar{L};\bar{\bar{L}}$$

$$\overline{L;L} \qquad \overline{\bar{L};L} \qquad \overline{L;\bar{L}} \qquad \overline{\bar{L};\bar{L}}.$$

Transitivity did not further reduce the list, though it did add some useful inclusions:

$$L;L\ ))\ L$$

$$L;\bar{L}\ ))\ \bar{L}$$

$$\bar{L};L\ ))\ \bar{L}.$$

If we now add that L is reflexive and invoke DR9.1 (or DR9.2), these inclusions are easily turned into equivalences.

DR10.1 $T(L), R(L) \vdash L;L = L$

DR10.2 $T(L), C(L), R(L) \vdash L;\bar{L} = \bar{L}$

DR10.3 $T(L), C(L), R(L) \vdash \bar{L};L = \bar{L}$

This means that when L is an equivalence relation, the eight logically distinct forms have been reduced to four:

$$L\ (= L;L,\ \overline{\bar{L};L},\ \overline{L;\bar{L}})$$

$$\bar{L}\ (= \overline{L;L},\ \bar{L};L,\ L;\bar{L})$$

$$\bar{L};\bar{L}$$

$$\overline{\bar{L};\bar{L}}.$$

For reasons like those mentioned at the end of section 8.6, this is as far as the reductions can go. When $x\bar{L};\bar{L}y$, x may either stand in L to y or not stand in L to y, depending upon the particular case. For instance, let equality of length be the equivalence relation in question. Then, from the fact that x is different in length from something that is different in length from y, it neither follows that x is different in length from y nor that it is the same length.

Equivalence relations will, of course, have other interesting properties. For instance,[6]

DR10.4. $T(L), C(L), R(L) \vdash L;(\bar{L}eM) = \bar{L}eM.$

Proof:		
	1. $L;L\ ))\ L$	Assumption
	2. $L\ ))\ L^{-1}$	Assumption
	3. $I\ ))\ L$	Assumption
	4. $I;(\bar{L}eM)\ ))\ L;(\bar{L}eM)$	3, R4r
	5. $\bar{L}eM\ ))\ L;(\bar{L}eM)$	4, Th9.5, RR

6. $L;(\bar{L}eM))(L;\bar{L})eM$ — Th8.2
7. $L;\bar{L}=\bar{L}$ — 1, 2, 3, DR10.2
8. $L;(\bar{L}eM))\bar{L}eM$ — 6, 7, RR
9. $L;(\bar{L}eM)=\bar{L}eM$ — 5, 8, R5i

With "I" available, other properties of relations can be defined. For instance, there is the *Existence Condition* which we introduced in section 8.6: A relation satisfies this condition in a universe if and only if every member of that universe stands in that relation to at least one member of that universe; i.e., when (x)(∃y)xLy. As we have seen, this will be so just in case $L;L^{-1}$ is reflexive. We thus define,

Df10.1. L satisfies the Existence Condition = df $I))L;L^{-1}$.

We will abbreviate this by "E(L)." It is then easy to prove,

DR10.5. R(L) ⊦ E(L).

Proof:
1. $I)L$ — Assumption
2. $I^{-1}))L^{-1}$ — 1, R3
3. $I))L^{-1}$ — 2, Th9.3, RR
4. $I;I))L;L^{-1}$ — 1, 3, DR7.1
5. $I))L;L^{-1}$ — 4, Th9.1, RR

It can also be shown that every transitive and convertible relation which satisfies the Existence Condition is reflexive.

DR10.6. T(L), C(L), E(L) ⊦ R(L)

We are now in a position to establish formally a claim which we made in section 8.5. In that section, we saw that De Morgan implicitly uses the Existence Condition in proving ?Th5.4. That proof uses the new principle that $L))L;(M;M^{-1})$, and De Morgan saw that this is closely related to the assumption that $M;M^{-1}$ is reflexive–that is, that M satisfies the Existence Condition. We can now show,

DR10.7. E(M) ⊦ $L))L;(M;M^{-1})$.

Proof:
1. $I))M;M^{-1}$ — Assumption
2. $L;I))L;(M;M^{-1})$ — 1, R4l
3. $L))L;(M;M^{-1})$ — 2, AI, RR

The implication in the other direction does not hold.

Other properties of relations can be expressed using the identity relation. The *Functionality Condition*, abbreviated by "F(L)," says that no element of the universe of discourse is related by L to more than one element of the universe.

> *Df10.2.* L satisfies the Functionality Condition = df $L^{-1};L\,))\,I$.

We can show that if L and M satisfy this condition, then (L;M) does.[7]

> *DR10.8.* F(L), F(M) ⊢ F(L;M)

Proof:		
1.	$L^{-1};L\,))\,I$	Assumption
2.	$M^{-1};M\,))\,I$	Assumption
3.	$M^{-1};(L^{-1};L)\,))\,M^{-1};I$	1, R4l
4.	$M^{-1};(L^{-1};L)\,))\,M^{-1}$	3, AI, RR
5.	$(M^{-1};(L^{-1};L));M\,))\,M^{-1};M$	4, R4r
6.	$(M^{-1};(L^{-1};L));M\,))\,I$	2, 5, R1
7.	$((M^{-1};L^{-1});L);M\,))\,I$	6, Th3.8, RR
8.	$(M^{-1};L^{-1});(L;M)\,))\,I$	7, Th3.8, RR
9.	$(L;M)^{-1};(L;M)\,))\,I$	8, A5, RR

We can now show that any relation which is reflexive and satisfies the Functionality Condition is identical to the identity relation.

> *DR10.9.* R(L), F(L) ⊢ L = I

Proof:		
1.	$I\,))\,L$	Assumption
2.	$L^{-1};L\,))\,I$	Assumption
3.	$I^{-1}\,))\,L^{-1}$	1, R3
4.	$I\,))\,L^{-1}$	3, Th9.3, RR
5.	$I;L\,))\,L^{-1};L$	4, R4r
6.	$L\,))\,L^{-1};L$	5, Th9.5, RR
7.	$L\,))\,I$	2, 6, R1
8.	$L = I$	1, 7, R5i

Theorems 9.1–9.4 allow us to show that I satisfies the Existence and Functionality Conditions.

> *Th10.1.* ⊢ $I\,))\,I;I^{-1}$

> *Th10.2.* ⊢ $I^{-1};I\,))\,I$

Finally, let us note an interesting fact. De Morgan had claimed that $L;L^{-1}$ is always reflexive, while we have seen that this is a material assumption, the Existence Condition, which some relations satisfy and others do not. Oddly enough, De Morgan seems not to have noticed that both LeL^{-1} and LoL^{-1} are reflexive, no matter what L is. For instance, "$x(LeL^{-1})x$" just means the same as "$(y)(xLy \supset xLy)$." The proof that $I)) LeL^{-1}$ requires several transformations.

Th10.3. $\vdash I)) LeL^{-1}$

Proof:	1. $(\bar{I}eL);L^{-1})) \bar{I}$	A6
	2. $\bar{\bar{I}})) \overline{(\bar{I}eL);L^{-1}}$	1, R2
	3. $I)) \overline{(\bar{I}eL);L^{-1}}$	2, A1, RR
	4. $I)) \overline{(\bar{I}eL)}eL^{-1}$	3, Th4.9, RR
	5. $I)) (\bar{\bar{I}};L)eL^{-1}$	4, Th4.11, RR
	6. $I)) (I;L)eL^{-1}$	5, A1, RR
	7. $I)) LeL^{-1}$	6, Th9.5, RR

This yields, almost immediately,

Th10.4. $\vdash I)) LoL^{-1}$.

From Th10.3, we can easily prove,

Th10.5. $\vdash Le(MeM^{-1}))) L$.

Proof.	1. $I)) MeM^{-1}$	Th10.3
	2. $Le(MeM^{-1}))) LeI$	1, DR7.4
	3. $Le(MeM^{-1}))) L$	2, Th9.6, RR

This theorem says, for instance, that if John loves every child of every parent of Mary, then John loves Mary. It is obviously valid and, since it contains only the symbols of D, it ought to be a theorem of D, not just of DI. Is it, in fact a theorem of D? Without firm metalogical theorems about the scope of D it is impossible to tell, but many failed attempts at deductions suggest that it may not be. We might, then, wish to add it as an additional axiom to D.

Principle 1: $Le(MeM^{-1}))) L$.

But there is another possibility. Since Th10.5 can be deduced within DI, and DI is of independent interest, we could use DI to formulate the logic of relations, rather than D alone. As noted above, this is the route taken by modern versions of the logic of relations.

Principle 1 is bothersome because it raises the question of whether there are other valid laws of relational logic that are not derivable in D. Such questions cannot be settled in the absence of metalogical results. But there are other valid laws which I have not been able to deduce in D. One example is,

Principle 2. $L)) (L;L^{-1});L$.

This principle can be used to show that any subset of the Identity relation shares two crucial features of that relation.[8]

DR10.10. $L)) I \vdash L = L^{-1}$

Proof:	1. $L)) I$	Assumption
	2. $(L;L^{-1});L)) (L;L^{-1});I$	1, R4l
	3. $(L;L^{-1});L)) L;L^{-1}$	2, AI, RR
	4. $L)) (L;L^{-1});L$	P2
	5. $L)) L;L^{-1}$	3, 4, R1
	6. $L^{-1})) I^{-1}$	1, R3
	7. $L^{-1})) I$	6, Th9.3, RR
	8. $L;L^{-1})) L;I$	7, R4l
	9. $L;L^{-1})) L$	8, AI, RR
	10. $L = L;L^{-1}$	5, 9, R5i
	11. $L^{-1} = (L;L^{-1})^{-1}$	10, DR3.10
	12. $L^{-1} = (L^{-1})^{-1};L^{-1}$	11, Th3.6, RR
	13. $L^{-1} = L;L^{-1}$	12, A2, RR
	14. $L = L^{-1}$	10, 13, RR

DR10.11. $L)) I \vdash L;L = L$

Proof.	1. $L)) I$	Assumption
	2. $L;L)) L;I$	1, R4l
	3. $L;L)) L$	2, AI, RR
	4. $(L;L);L)) L;L$	3, R4r
	5. $L)) (L;L^{-1});L$	P2
	6. $L)) (L;L);L$	5, DR10.10, RR
	7. $L)) L;L$	6, 4, R1
	8. $L;L = L$	3, 7, R5i

When $L)) I$, L has some other properties of I which will be useful in section 12. These include,

DR10.12 $L)) I \vdash (M;L);N)) M;N$

DR10.13 $L;O)) I \vdash (M;L);(O;N))) M;N$.

Notice that these conditions apply to relations which satisfy the Functionality Condition–i.e., are such that $L^{-1};L)) I$. Thus,

DR10.14 $F(L) \vdash (M;L^{-1});(L;N))) M;N$.

Finally, let us note that P2 also allows us to prove that when L is convertible, the inclusion that constitutes the definition of transitivity can be strengthened to an equivalence.

DR10.15 $C(L), T(L) \vdash L;L = L$

Proof.		
	1. $L)) L^{-1}$	Assumption
	2. $L;L)) L$	Assumption
	3. $(L;L);L)) L;L$	2, R4r
	4. $L = L^{-1}$	1, DR6.21
	5. $(L;L^{-1});L)) L;L$	3, 4, RR
	6. $L)) (L;L^{-1});L$	P2
	7. $L)) L;L$	5, 6, RR
	8. $L;L = L$	2, 7, R5i

11. A SURROGATE FOR QUANTIFICATION THEORY

De Morgan's theory of the syllogism is well-known for the fact that its term letters cannot be taken to stand for either the null class or the universal class. In general, both X and non-X must have members. This condition is not quite as stringent as it seems, for De Morgan allows that the universe of discourse need not be restricted to actual objects. But the principle stands: given a universe of discourse, it must contain both Xs and non-Xs, for every term letter "X."

At the same time, it is not difficult to formulate in De Morgan's syllogistic language a proposition which states that X is the null set: it is just "X)) non-X," which translates as "all X are non-X." On the Boolean interpretation of categorical propositions, this is equivalent to "nothing is X." Similarly, the claim that everything is X, that "all non-X are X," is expressed by "non-X)) X."

This way of stating that classes are null or universal would be precluded in De Morgan's syllogistic language, as it is in Aristotle's, by restricting categorical propositions to those in which the subject and predicate are completely different, so that one cannot be the complement of the other.[9]

We have followed De Morgan's general point of view on this issue in constructing D and DI, where we use relation terms instead of class terms. The language contains no symbols for the null and universal relations. Our decision to do so was reinforced by the fact that De Morgan never introduced such terms. Yet our language provides an easy way to assert that L is a null or a universal relation.

Df11.1. L is a null relation = df $L\,))\,\bar{L}$.

Df11.2. L is a universal relation = df $\bar{L}\,))\,L$.

We will abbreviate these by "N(L)" and "U(L)," respectively.

The situation here is parallel to that in syllogistic logic, except that there is nothing in our formation rules to block treating $L\,))\,\bar{L}$ and $\bar{L}\,))\,L$ as sentences. Such a rule could have been added, but it flies in the face of the fact that $L\,))\,L$ is repeatedly used as a sentence of D. This makes the case quite different than the syllogistic one, and it would be very artificial to exclude $L\,))\,\bar{L}$ while including $L\,))\,L$.

Once we allow such conditions, some interesting results follow.

DR11.1. $L\,))\,M,\ L\,))\,\bar{M} \vdash N(L)$

Proof:		
1.	$L\,))\,M$	Assumption
2.	$L\,))\,\bar{M}$	Assumption
3.	$\bar{M}\,))\,\bar{L}$	1, R2
4.	$L\,))\,\bar{L}$	2, 3, R1

That is, only the null relation is included in both M and $\bar{M}$. Similarly, only the universal relation includes both M and $\bar{M}$.

DR11.2. $M\,))\,L,\ \bar{M}\,))\,L \vdash U(L)$

On the interpretation we have been pursuing, the null relation should be included in every relation. This appears to require adding a new principle to D, which we state as,

Principle 3. From N(L) infer $L\,))\,M$.

From this it follows that every relation is contained in the universal relation.

DR11.3. $U(L) \vdash M\,))\,L$

We can use P3 to prove that there is only one null relation.

DR11.4. N(L), N(M) ⊢ L = M

Proof:	1. $L)) \bar{L}$	Assumption
	2. $M)) \bar{M}$	Assumption
	3. L)) M	1, P3
	4. M)) L	2, P3
	5. L = M	3, 4, R5i

We can also show that there is only one universal relation.

DR11.5. U(L), U(M) ⊢ L = M

At this point, we will shift gears and approach U(L) from another point of view. From this standpoint, we want to know if it is possible to construct a surrogate for a part of quantification theory within De Morgan's logic. This is the part consisting of doubly-quantified propositions. U(L) already gives us one such form:

In Standard Logic	*In DI*
1. (x)(y)xLY	$\bar{L})) L.$

We now want to see how all the other doubly-quantified forms can be written. Changing the order of quantifiers is easy enough.

2. (y)(x)xLy	$\overline{L^{-1}})) L^{-1}$

These are the only forms that can be expressed in D; but DI gives two more. We have already seen one of these in our discussion of the Existence Condition; the other reverses the order of the variables.

3. (x)(∃y)xLy	$I)) L;L^{-1}$
4. (y)(∃x)xLy	$I)) L^{-1};L$

The notation of DI has allowed us to express the four doubly-quantified forms which begin with the universal quantifier. But it will not allow us to express any forms which begin with an existential quantifier. The reason for this is simple: existential statements are the negations of universally quantified statements, so that they would have to be expressed in DI through negated inclusion statements. But DI has no propositional connectives at all, not even negation.

De Morgan's logic of relations is not alone in not having negation. Aristotle's logic is best formulated in a system without propositional negation. Instead, the four forms of categorical propositions are given independently, and no negation sign occurs at all. Its place is taken by the stipulation that A and O propositions are contradictories of each other, as are E and I propositions. Common modes of reasoning are then formalized with a *reductio* rule for cases in which both a proposition and its contradictory are generated.[10]

It is also well known that Boole's logic did not allow class identities to be negated. In the absence of this, he had to express the particular proposition, "some x are y," by the form "x = vy," where v is the "elective" symbol whose function is to generate an unspecified subclass of y.

There is much to be said for introducing into DI some means for negating inclusions. It would not only make possible the completion of our system of quantification, but it would increase the expressive power of DI in other ways as well. If we have a way to say that a relation is null or universal, it seems only natural that we also have a way to say that it is non-null or non-universal. There are at least two ways in which we could do this. One would be to use the Aristotelian model. We could introduce a third and fourth binary relation, non-inclusion and non-identity, which we could symbolize as, say, "))" and "≠." We could then define contradictoriness in the obvious way and provide a *reductio* rule for handling cases where two contradictory propositions are generated. The other method would be to introduce propositional negation, along with a new formation rule for it and logical rules for using it (e.g., negation introduction and negation elimination).

It is not immediately clear which of these methods is the best one. Since it would be a genuine addition to De Morgan's logic of relations, there is nothing in that logic to serve as a guide. If we attempt to make this harmonize with De Morgan's general viewpoint about categorical propositions, though, we do get some guidance.

In section 3.2, we encountered the following symbolization:

All X are Y	X))Y
Some X are not Y	X(.(Y.

Parentheses indicate whether a term is being used universally or particularly. Lack of a dot means that the proposition is affirmative, and one dot means that it is negative.

The interpretation of this notation is a bit unclear because De Morgan's "))" and "(.(" are complex symbols, each element of which has a meaning. But one thing is clear: there is no sign for propositional negation. The dot means that the proposition is a negative proposition, but this does not mean that the dot is a sign of propositional negation. This suggests that we use signs of non-inclusion and non-identity in augmenting DI, rather than adding a sign for propositional negation. This has the added advantage of avoiding even a hint of the larger propositional logic in reconstructing and now augmenting De Morgan's logic.

Since we are not transposing De Morgan's complex apparatus for categorical propositions to the relational level, we need not use such internally complex forms as "(.(" to represent non-inclusion. We will merely use "$\not))$". Also, since our concern is more with sketching the expressive power of the symbolism than with complex derivations, we will not specify the precise *reductio* rules for operating with this symbolism.

Let us now see how the other doubly-quantified propositions can be expressed in this system, which we will call DIN. They are,

	In Standard Logic	*In DIN*
5.	$(\exists x)(\exists y)xLy$	$L \not)) \bar{L}$
6.	$(\exists y)(\exists x)xLy$	$L^{-1} \not)) \overline{L^{-1}}$
7.	$(\exists x)(y)xLy$	$I \not)) \overline{\bar{L};L^{-1}}$
8.	$(\exists y)(x)xLy$	$I \not)) \overline{L^{-1}};L.$

The logical relationships among these 8 propositions, expressed in quantificational form, can be shown by the following chart:

$(x)(y)xLy$	1.	⟷	2.	$(y)(x)xLy$
	↓		↓	
$(\exists x)(y)xLy$	7.		8.	$(\exists y)(x)xLy$
	↓		↓	
$(y)(\exists x)xLy$	4.		3.	$(x)(\exists y)xLy$
	↓		↓	
$(\exists y)(\exists x)xLy$	6.	⟷	5.	$(\exists x)(\exists y)xLy$

We will not go through the tedious exercise of demonstrating all these logical relationships in DIN, contenting ourselves with stating the basic strategy needed and the two principles which must be assumed. The basic idea is to demonstrate the eight relationships along the perimeter of the diagram, and then to use the transivity of derivability to show the rest. The number of primitive logical relations along the perimeter can be reduced to two. The equivalence of 1 and 2 can be shown easily using Rule 3 and Th3.4, as can that of 6 and 5. The right-hand column is just like the left-hand column, with L^{-1} replacing L. The inference from 4 to 6 is merely a substitution instance of the inference from 1 to 7, with L^{-1} replacing L; and similarly, the move from 3 to 5 is an instance of that from 2 to 8. This means that only two new principles are needed for all these logical relationships:

Principle 4: From $\bar{\mathrm{L}}\,))\,\mathrm{L}$ infer $\mathrm{I}\not{))}\,\bar{\mathrm{L}};\overline{\mathrm{L}^{-1}}$. $(1 \vdash 7)$
Principle 5: From $\mathrm{I}\not{))}\,\bar{\mathrm{L}};\mathrm{L}^{-1}$ infer $\mathrm{I}\,))\,\mathrm{L}^{-1};\mathrm{L}$. $(7 \vdash 4)$

It must be admitted that P4 and P5 lack intuitive appeal; and their main justification lies in the fact that they give us the quantificational inferences we need. In the language of DIN, both of them involve an inclusion and a non-inclusion. Our intuitive grasp of them is likely to be found more in our understanding of their quantificational translations than in a sense of the forms themselves. Can anything more be said to motivate them, other than the fact that they work?

It would be reassuring if we could either deduce P4 or P5 from simpler principles or motivate them in a way which is not totally dependent upon standard logic. I know of no way to deduce these principles from simpler ones; but something more can be said about their intuitive significance. P4 has a fairly clear intuitive meaning in quantificational terms, since it is merely an inference from a universally quantified sentence to an existentially quantified one. But P5 is more opaque. It is equivalent to saying that if $\bar{\mathrm{L}}$ does not satisfy the Existence Condition then L^{-1} does. It is hard to see why this is so without engaging in explicitly quantificational reasoning: if it is not the case that everything bears $\bar{\mathrm{L}}$ to something, then something does not bear $\bar{\mathrm{L}}$ to anything; that is, it bears L to everything. So, for everything, there is something that bears L to it, and thus everything bears L^{-1} to something. This means that L^{-1} satisfies the Existence Condition. None of this goes very far, though, in explaining P5 purely in terms of DIN.

These principles are related to the assumption that the universe of discourse is not empty. Marching around the perimeter of the diagram, we can prove that if L is the universal relation then it is non-null.

DR11.6. $\bar{L}))L \vdash L\not)\not)\bar{L}$

This is an interesting result because it does not use the identity relation. We could have developed the topics in this section in a different order, introducing non-inclusion before identity; and in this order of development, DR 11.6 would have been a candidate for a new principle for the enlarged system. An instance of DR 11.6 is the result that if L is the null relation, then $\bar{L}$ is non-null.

DR11.7. $L))\bar{L} \vdash \bar{L}\not)\not)\bar{\bar{L}}$

If we think of relations as sets of ordered pairs, this means that there is at least one ordered pair.

A straightforward existence assertion can be proved within DIN.

Th11.1. $\vdash I\not)\not)\bar{I}$

Proof:	1. $I))I;I^{-1}$	Th10.1
	2. $I\not)\not)\bar{I}$	1, instance of form 3 implies form 5.

This is not a mere implication, like DR11.7, but a simple assertion. Since it says the same thing as "(∃x)(∃y)xIy," it asserts that *something* exists.

It must be admitted that we can interpret this as an existence assertion only because we have adopted a certain interpretation of the sentences of DIN, with their corresponding translation into standard logic. It is one which turns non-inclusion statements into existence assertions. This seems consistent with De Morgan's viewpoint; and the mapping between sentences of DIN and standard logic is preserved by it.

12. POSTSCRIPT – 1864

Some four years after publishing S4, De Morgan added the following annotation to his copy of it:

13 July 1864. When M^{-1} has only one M, write it $\dot{M}^{-1}$. Then $M\dot{M}^{-1}X$ is X, and $\dot{M}^{-1}MX$ may be other than X, but is transitive. For $\dot{M}^{-1}M\dot{M}^{-1}MX$ is $\dot{M}^{-1}(M\dot{M}^{-1})MX$, or $\dot{M}^{-1}MX$. Are all transitive and convertible relations reducible to $\dot{M}^{-1}M$? (S4, 236).

This short annotation contains several intriguing suggestions. Our task will be to interpret it in an augmented version of DI in such a way as to yield the results which De Morgan claimed and, in the process, generate other results as well.

Interpreting De Morgan's notation here is made difficult by the fact that he treats "$\dot{M}^{-1}$" as a unit, while rigor suggests that the dot be separated from the converse-sign. Once this separation is made, two questions arise:

1. What does the dot-operator mean?
2. How is "$\dot{M}^{-1}$" to be parenthesized? Does it mean $(\dot{M})^{-1}$, or does it mean $(M^{-1})^{\cdot}$?

It turns out that these two questions are related, since the answer to the first dictates the answer to the second. We will consider them in turn.

The dot-operator is clearly a uniqueness operator, but it is less clear just what is taken to be unique. The sentence "a$\dot{M}$b" could have two readings:

$$\begin{aligned}(1)\ a\dot{M}b &\equiv \text{a is the only M of b}\\ &\equiv aMb\ \&\ (x)(xMb \supset a = x)\\ &\equiv aMb\ \&\ a(IeM)b;\\ (2)\ a\dot{M}b &\equiv \text{a is an M only of b}\\ &\equiv aMb\ \&\ (x)(aMx \supset x = b)\\ &\equiv aMb\ \&\ a(MoI)b.\end{aligned}$$

This second reading is equivalent to,

$$\begin{aligned}a\dot{M}b &\equiv \text{b is the only } M^{-1} \text{ of a}\\ &\equiv bM^{-1}a\ \&\ (x)(xM^{-1}a \supset b = x)\\ &\equiv bM^{-1}a\ \&\ b(IeM^{-1})a.\end{aligned}$$

Thus, if we let aMb = a is a child of b, the first reading is,

$$\text{a child b} \equiv \text{a is the only child of b},$$

while the second reading is,

$$\begin{aligned}\text{a child b} &\equiv \text{a is a child only of b}\\ &\equiv \text{b is the only parent of a.}\end{aligned}$$

We will soon see that there is no reason to prefer one reading over the other. On grounds solely of linguistic ease, I will *adopt the first reading* and introduce a new symbol, "$\overset{\times}{M}$," for the second reading. This can be defined,

$$a\overset{\times}{M}b = \text{df } b(M^{-1})^{\cdot} a.$$

The second notational question concerns the placement of parentheses for $\dot{M}^{-1}$. The two options have quite different meanings:

$$\begin{aligned} a(\dot{M})^{-1}b &\equiv b\dot{M}a \\ &\equiv bMA \ \& \ (x)(xMa \supset x = b) \\ &\equiv bMa \ \& \ b(IeM)a; \\ a(M^{-1})^{\cdot}b &\equiv aM^{-1}b \ \& \ (x)(xM^{-1}b \supset x = a) \\ &\equiv bMa \ \& \ (x)(bMx \supset x = a) \\ &\equiv bMa \ \& \ b(MoI)a. \end{aligned}$$

Which of these should be selected as the reading for "a$\dot{M}^{-1}$b"? De Morgan's account of "$\dot{M}^{-1}$" says, "When M^{-1} has only one M, write it $\dot{M}^{-1}$." If M = child, this means, "when parent has only one child, write it $\dot{\text{child}}^{-1}$." This suggests that "a $\dot{\text{child}}^{-1}$b" means that a has only one child, which is b; that is, that b is the only child of a. This is just what the first reading gives; and we will see that this also gives the results which De Morgan claims for M;$\dot{M}^{-1}$ and $\dot{M}^{-1}$;M.

While we are at it, we will parenthesize $\overset{\times}{M}^{-1}$ as well. These results, though, are equivalent to what we have already obtained:

$$\begin{aligned} a(\overset{\times}{M})^{-1}b &\equiv b\overset{\times}{M}a \equiv a(M^{-1})^{\cdot}b; \text{ and,} \\ a(M^{-1})^{\times}b &\equiv b((M^{-1})^{-1})^{\cdot} a \equiv b\dot{M}a \equiv a(\dot{M})^{-1}b. \end{aligned}$$

This explains why there was no clear reason for choosing our "·" over "×" as the reading for De Morgan's original dot notation: they are easily intertranslatable, when care is given to the placement of parentheses. Thus, De Morgan's "$\dot{M}^{-1}$" can be represented by either "$(\dot{M})^{-1}$" or "$(M^{-1})^{\times}$." We will emphasize the former.

Let us now augment our system DI by adding the dot operator, and saying that if L is a relation term, then $(L)^{\cdot}$ is also a relation term. (When convenient we will drop parentheses and write $\dot{L}^{-1}$ for $(\dot{L})^{-1}$.) What axioms or rules should characterize $\dot{L}$? We have seen that the basic equivalence is,

$$a\dot{L}b \equiv aLb \ \& \ a(IeL)b.$$

If we had Boolean multiplication available in DI, we could represent this by,

$$\dot{L} = L \cdot (\mathrm{IeL}).$$

Without this Boolean operation, though, we can get much the same effect by adopting two new axioms:

$$\text{AD1.}\ \dot{L}\,))\,L$$
$$\text{AD2.}\ \dot{L}\,))\,\mathrm{IeL}.$$

In generating De Morgan's results, it will be convenient to begin with some preliminary theorems.

Th12.1. $\vdash \dot{L};L^{-1}\,))\,I$

Proof:		
	1. $\dot{L}\,))\,\mathrm{IeL}$	AD2
	2. $\dot{L};L^{-1}\,))\,(\mathrm{IeL});L^{-1}$	1, R4r
	3. $(\mathrm{IeL});L^{-1}\,))\,I$	A6
	4. $\dot{L};L^{-1}\,))\,I$	2, 3, R1

When we apply DR10.13 to Th12.1, we obtain,

Th12.2. $\vdash (L^{-1};\dot{L});(L^{-1};\dot{L})\,))\,L^{-1};\dot{L}.$

In section 8.10, we saw that any subset of the identity relation possesses a strong form of transitivity and is also convertible. DR10.10 and DR10.11 allow us to prove, respectively,

Th12.3. $\vdash (\dot{L};L^{-1});(\dot{L};L^{-1}) = (\dot{L};L^{-1})$

Th12.4. $\vdash \dot{L};L^{-1} = L;\dot{L}^{-1}.$

The equivalence in Th12.4 then allows us to prove variants of Th12.1–Th12.3. They are, in order,

Th12.5. $\vdash L;\dot{L}^{-1}\,))\,I$

Th12.6. $\vdash (\dot{L}^{-1};L);(\dot{L}^{-1};L)\,))\,\dot{L}^{-1};L$

Th12.7. $\vdash (L;\dot{L}^{-1});(L;\dot{L}^{-1}) = L;\dot{L}^{-1}.$

We have now proved, in Th12.5 and Th12.6, the two positive results which De Morgan stated in his remark. It is interesting to note that the proof of Th12.1, on which that of Th12.5 depends, utilized not only axiom AD2, but axiom A6 (which was essential to the proof of Theorem K) as well. Also, De Morgan is correct in seeing that the proof of the transitivity of $\dot{L}^{-1};L$ (Th12.6) requires the associativity of ";", for it is used in proving DR10.13 on which Th12.2 (and thus

Th12.6) is based. This is one of the very few occasions on which De Morgan noted the need for this principle in his logic of relations, despite his recognition of it in algebra.

De Morgan also notes that $\dot{L}^{-1}; L$ is not included in I. The truth of this can be seen by noting that a $\dot{L}^{-1}$;Lb is equivalent to $(\exists z)(z\dot{L}a$ & zLb), from which it does not follow that aIb. (This contrasts with $aL;\dot{L}^{-1}b$, which is equivalent to $(\exists z)(aLz$ & $b\dot{L}z)$, and which does imply aIb.) An interesting inclusion involving $\dot{L}^{-1}$;L can still be proved. We begin with,

Th12.8. $\vdash L^{-1};\dot{L})) L^{-1}eL.$

Proof:		
1.	$\dot{L})) IeL$	AD2
2.	$L^{-1};\dot{L})) L^{-1};(IeL)$	1, R4l
3.	$L^{-1};(IeL))) (L^{-1};I)eL$	Th8.2
4.	$L^{-1};\dot{L})) (L^{-1};I)eL$	2, 3, R1
5.	$L^{-1};\dot{L})) L^{-1}eL$	4, AI, RR

By taking the converse of each side of this inclusion, we obtain,

Th12.9. $\vdash \dot{L}^{-1};L)) L^{-1}oL.$

Thus, while $\dot{L}^{-1}$;L is not included in I, it is included in $L^{-1}oL$.

The ×-operator is easily defined when we remember that $a\overset{\times}{L}b$ is equivalent to $b(L^{-1})^{\cdot}\, a$. Thus we can define,

*Df*12.1. $\overset{\times}{L} = \text{df } ((L^{-1})^{\cdot})^{-1}.$

From this definition we can derive the two theorems which are analogous to the axioms for the dot-operator.

Th12.10 $\vdash \overset{\times}{L})) L$

Proof:		
1.	$(L^{-1})^{\cdot})) L^{-1}$	AD1
2.	$((L^{-1})^{\cdot})^{-1})) (L^{-1})^{-1}$	1, R3
3.	$\overset{\times}{L})) (L^{-1})^{-1}$	2, D12.1, RR
4.	$\overset{\times}{L})) L$	3, A2, RR.

Th12.11. $\vdash \overset{\times}{L})) LoI$

Proof:		
1.	$(L^{-1})^{\cdot})) IeL^{-1}$	AD2
2.	$((L^{-1})^{\cdot})^{-1})) (IeL^{-1})^{-1}$	1, R3
3.	$\overset{\times}{L})) (IeL^{-1})^{-1}$	2, D12.1, RR

4. $\overset{\times}{L}$)) $(L^{-1})^{-1}oI^{-1}$ 3, Th4.6, RR
5. $\overset{\times}{L}$)) LoI^{-1} 4, A2, RR
6. $\overset{\times}{L}$)) LoI 5, Th9.3, RR

Finally, let us note the following equivalences which we have already mentioned.

Th12.12. $\vdash (\overset{\times}{L})^{-1} = (L^{-1})^{\cdot}$

Th12.13. $\vdash (\dot{L})^{-1} = (L^{-1})^{\times}$.

13. DE MORGAN'S CONJECTURES

In considering the properties of relations, we have seen that De Morgan proposed two conjectures concerning the reducibility of a relation to the relative product of other relations. Those conjectures are,

From 1860: Every convertible relation can be reduced to the form $M;M^{-1}$.

From 1864: Every convertible and transitive relation can be reduced to the form $\dot{M}^{-1};M$.

In this section we will discuss these two conjectures along with three others, arranging them in a natural order.

Conjecture 1 (1860): Every convertible relation can be reduced to the form $M;M^{-1}$. We can show the logical form of this conjecture by expressing it in a symbolic form.

C1: $C(L) \supset (\exists M)(x)(y)[xLy \equiv xM;M^{-1}y]$.

That is,

$C(L) \supset (\exists M)(x)(y)[xLy \equiv (\exists z)(xMz \;\&\; yMz)]$.

We have already seen that this conjecture is false: every relation which is reducible to $M;M^{-1}$ is weakly reflexive; that is, for such an L, $(x)(y)[xLy \supset (xLx \;\&\; yLy)]$. But there are convertible relations, such as adjacency (expressed by "next to"), which are convertible but not weakly reflexive. Thus, we saw that C1 had to be weakened to read,

Conjecture 2: Every convertible and weakly reflexive relation can be reduced to the form $M;M^{-1}$.

That is,

C2: $[C(L) \;\&\; WR(L)] \supset (\exists M)(x)(y)[xLy \equiv (\exists z)(xMz \;\&\; yMz)]$.

We can, in De Morgan's words, leave this as a conjecture whose "material truth" remains to be decided and ask, instead, whether C2 might be modified in interesting ways. While De Morgan refers to reflexivity (without calling it that), he does not consider weak reflexivity; and we have seen that weak reflexivity is not expressible in DI. He does, however, deal with transitivity at length, and a weakened version of C2 can be motivated by noting that the following principle can be proved easily in standard logic:

Every convertible and transitive relation is weakly reflexive.

This principle and C2 then imply,

Conjecture 3: Every convertible and transitive relation can be reduced to the form $M;M^{-1}$.

We express this as,

$$C3: \; [C(L) \; \& \; T(L)] \supset (\exists M)(x)(y)[xLy \equiv (\exists z)(xMz \; \& \; yMz)].$$

C3 is weaker than C2, since C3 is implied by C2 but not vice versa.

This conjecture does not seem to have the same problems which either of the first two conjectures faced. It does, however, have a problem of its own. Since L is both convertible and transitive, and since L is identical to $M;M^{-1}$, the latter should also be convertible and transitive.

This means that the M which we select for this reduction must be such that the resulting $M;M^{-1}$ is convertible and transitive. It turns out that this will depend upon the specific M which we select, since $M;M^{-1}$ is not always transitive. This is so even for obvious candidates for reduction. Let L, for instance, be the relation of *sibling*, expanded to include half siblings as well as full siblings. Then the natural M is just the relation *is a child of*. If we let "S" and "C" stand for these two relations, we have,

$$\begin{aligned} aSb &\equiv (\exists w)(aCw \; \& \; wC^{-1}b) \\ &\equiv (\exists w)(aCw \; \& \; bCw). \end{aligned}$$

In this case, of course, the non-transitivity of $C; C^{-1}$ merely reflects that of S, so there is no problem. But when the L is transitive, we would like the conjecture framed in a firmer way to take the needed transitivity into account.

De Morgan's method for approaching this problem is suggested by looking at the context for his first conjecture. He was led to conjecture that every convertible L is reducible to some $M;M^{-1}$ *after* he had noted that $M;M^{-1}$ is always convertible. By parity of form, we should conjecture that every convertible and transitive L is reducible to some $M;M^{-1}$ *only when* we are assured that $M;M^{-1}$ is always convertible and transitive. But we have just seen that, while $M;M^{-1}$ is always convertible, it is not always transitive. If we are to get a viable reduction, some other function of M must be found.

De Morgan thinks (in 1864) that he may have found such a function in $\dot{M}^{-1};M$. Following the methodology of Conjecture 1, he notes first that $\dot{M}^{-1};M$ is transitive; and, apparently assuming that it is also convertible (like $M;M^{-1}$), suggests:

Conjecture 4: Every convertible and transitive relation can be reduced to the form $\dot{M}^{-1};M$.

That is,

$$C4\colon [C(L) \,\&\, T(L)] \supset (\exists M)(x)(y)[xLy \equiv (\exists z)(x\dot{M}^{-1}z \,\&\, zMy)],$$

which is the same as,

$$[C(L) \,\&\, T(L)] \supset (\exists M)(x)(y)[xLy \equiv (\exists z)(z\dot{M}x \,\&\, zMy)].$$

Unfortunately, while $\dot{M}^{-1};M$ is transitive, it is not convertible. This can be seen by noting that,

$$a\dot{M}^{-1};Mb \equiv (\exists z)(z\dot{M}a \,\&\, zMb),$$

while the converse of $\dot{M}^{-1};M$ is $M^{-1};\dot{M}$, yielding,

$$\begin{aligned} a(\dot{M}^{-1};M)^{-1}b &\equiv aM^{-1};\dot{M}b \\ &\equiv (\exists z)(zMa \,\&\, z\dot{M}b). \end{aligned}$$

Since the uniqueness clauses are singled out differently in the two cases, the two forms are not equivalent. Thus, De Morgan is mistaken in assuming that $\dot{M}^{-1};M$ is convertible.

Can we find some other composite relation term which will be both transitive and convertible, and thus suitable for a more accurate conjecture? The theorems from section 8.12 show that this is possible. Theorems 12.4 and 12.7 convince us that De Morgan should have reversed the order of composition of his terms, since they show that $M;\dot{M}^{-1}$ is both convertible and transitive. This is due, in turn, to the fact that $M;\dot{M}^{-1}$ is included in I. Our final conjecture, then, is,

Conjecture 5: Every convertible and transitive relation can be reduced to the form $M;\dot{M}^{-1}$.

That is,

$$C5\text{: } [C(L) \,\&\, T(L)] \supset (\exists M)(x)(y)[xLy \equiv (\exists z)(xMz \,\&\, z\dot{M}^{-1}y)].$$

which is,

$$[C(L) \,\&\, T(L)] \supset (\exists M)(x)(y)[xLy \equiv (\exists z)(xMz \,\&\, y\dot{M}z)].$$

(It should be noted that the convertibility and transitivity of L imply, by DR10.15, that L;L = L; thus, the fact that Th12.7 is an equivalence and not an inclusion presents no problem for this conjecture.)

Conjecture 5 seems formally satisfactory. But is it true? It turns out that this conjecture is implied by the Axiom of Choice.

To see why this is so, note first that the class of individuals which constitutes our universe of discourse will contain two kinds of elements: those which are in the field of L (i.e., either in the domain or range of L) and those which are not. We need deal only with the former type.

By hypothesis, L is convertible and transitive. As we have noted, this means that it is also weakly reflexive; that is,

$$(x)(y)[xLy \supset (xLx \,\&\, yLy)].$$

But this implies that L is reflexive in the field of L. This means that if we limit our attention to those elements of the universe of discourse which are in the field of L, then L is an equivalence relation.

It is well known that an equivalence relation, L, partitions a set into mutually exclusive subsets which are such that every member of a subset stands in L to every member of that subset and does not stand in L to any member of any other of these subsets. This means that *a* will stand in L to *b* if and only if they are members of the same subset under this partition. Thus, the relation L has been reduced to common membership in an equivalence class generated by L.

This would suffice to prove Conjecture 5 were it not for one fact: the equivalence classes which are used are not members of the universe of discourse, since De Morgan clearly assumes that the universe will contain only individuals. Reducibility, for De Morgan, requires a common relationship to some *individual*, not to an equivalence class. Where is this individual to come from?

Here is where the Axiom of Choice enters. If we assume this axiom, then for each equivalence class, B, we can select a member, F(B), to "stand in" for the equivalence class. Common membership in B can now be replaced by a common relationship to F(B). Then aLb if and only if both *a* and *b* stand in this relation to some *z* which is the chosen object in the equivalence class to which *a* and *b* belong. The uniqueness condition in C5 will be guaranteed by the fact that the equivalence class under L for any element is always unique and that the chosen object within that class is unique.

Thus, C5 follows from the Axiom of Choice. It is not clear whether the converse implication holds: that is, whether the Axiom of Choice follows from C5. Whether or not this is the case, though, we can note with admiration the fact that De Morgan has raised an issue of surprising depth.[11]

NOTES

NOTE ON ABBREVIATIONS

For ease of reference, the following abbreviations will be used in referring to De Morgan's major logical works. These references will usually be inserted in the text.

SDM *On the Study and Difficulty of Mathematics* (1831; reprint, La Salle, Illinois: Open Court Publishing Company, 1943).

FL *Formal Logic: or, The Calculus of Inference, Necessary and Probable* (1847; reprint, ed. A.E. Taylor, London: The Open Court Company, 1926). All page references will be to the reprint pagination.

OS *On the Syllogism and Other Logical Writings*, ed. Peter Heath (New Haven: Yale University Press, 1966). Specific items from this volume will be referred to as follows:

S1 "On the Structure of the Syllogism" (1846), 1–21.

S2 "On the Symbols of Logic, the Theory of the Syllogism, and in particular of the Copula" (1850), 22–68.

S3 "On the Syllogism: III and on Logic in general" (1858), 74–146.

S *Syllabus of a Proposed System of Logic* (1860), 147–207.

S4 "On the Syllogism: IV and on the Logic of Relations" (1860), 208–46.

L "Logic," in *English Cyclopaedia, Arts and Sciences*, Vol. V (1860), 247–70.

S5 "On the Syllogism: V and on various points of the Onymatic System" (1863), 271–345.

All page references to these writings will be to the *OS* edition.

CHAPTER I

[1] S4, 237; L, 247.
[2] References will be to Richard Whately, *Elements of Logic* (New York: William Jackson/Boston: James Munroe & Co., 1836). This is stated to be "from the last London edition." Two very useful recent studies of Whately in his historical context are Calvin Jongsma, *Richard Whately and the Revival of Syllogistic Logic in Great Britain in the Early Nineteenth Century* (diss., University of Toronto, 1982) and James van Evra, "Richard Whately and the Rise of Modern Logic," *History and Philosophy of Logic* 5 (1984): 1-18.
[3] *Encyclopedia of Philosophy*, *s.v.* "Whately, Richard."

[4] *Logic*, 7.
[5] *Logic*, 146.
[6] *Logic*, 86.
[7] *Logic*, 46–47.
[8] *Logic*, 47.
[9] *Logic*, 48. He later brings in the several traditional rules for the syllogism, such as that of Undistributed Middle, to avoid the "occasional tediousness" of always using the *dictum*.
[10] *Logic*, 50.
[11] *Logic*, 53.
[12] *Logic*, 54.
[13] *Logic*, 86.
[14] *Logic*, 116. Boole also uses this type of reduction. De Morgan uses a similar reduction, but takes the first premise to be "Every case of p is a case of q." See FL, 25.
[15] *Logic*, 209.
[16] Ian Mueller, "Greek Mathematics and Greek Logic," in *Ancient Logic and its Modern Interpretations*, ed. John Corcoran (Dordrecht: D. Reidel Publishing Co., 1974), 35–70.
[17] On this point see Wesley C. Salmon, "In Praise of Relevance," *Teaching Philosophy* 4 (1981): 261.
[18] T.L. Heath, *The Thirteen Books of Euclid's Elements*, Vol. I, 2d ed. (1925; reprint, New York: Dover Publications, 1956), 241–42.
[19] Sir William Hamilton, "On the Study of Mathematics as an Exercise of Mind" (1836), in *Discussions on Philosophy and Literature* (New York: Harper & Bros., 1853), 279. Hamilton also refers to this attempt in his edition of *The Works of Thomas Reid*, Vol. II, 6th ed. (Edinburgh: Maclachlan and Stewart, 1863), 702n. H.L. Mansel inserts one of these demonstrations in the third edition of his *Artis Logicae Rudimenta, from the Text of Aldrich* (Oxford, 1856), 264–66.
[20] Christian Herlinus and Cunradus Dasypodius, *Analyseis Geometricae Sex Liborum Euclidis* (Strasbourg: J. Rihel, 1566). The translations which follow are by Nathan Greenberg, Professor of Classics, Oberlin College. For more information on Dasypodius see *Dictionary of Scientific Biography*, *s.v.* "Dasypodius, Cunradus"; and Miriam Usher Chrisman, *Lay Culture, Learned Culture: Books and Social Change in Strasbourg, 1480–1599* (New Haven and London: Yale University Press, 1982).
[21] *Analyseis*, "Preface."
[22] *Analyseis*, "Preface."
[23] *Analyseis*, II.
[24] Gottfried Wilhelm Leibniz, *Philosophical Papers and Letters*, Vol. I, ed. Leroy Loemker (Chicago: University of Chicago Press, 1956), 453. In this quotation, "sound" seems to mean the same as "valid," rather than "valid with true premises."
[25] "Greek Mathematics," 40–42.
[26] *The Works of Thomas Reid*, Vol. II, 631–714.
[27] *Brief Account*, 702.
[28] *Brief Account*, 702n.
[29] "Greek Mathematics," 42.
[30] In Note C, "On Material and Formal Consequence," in *Artis Logicae Rudimenta, from the Text of Aldrich* (Oxford: William Graham, 1849), App. 31–32.

[31] *Artis Logicae*, App. 30–31.
[32] *Brief Account*, 697n.
[33] *Artis Logicae*, App. 31.
[34] *Artis Logicae*, App. 32.
[35] In Note H, "On the Demonstrative Syllogism," *Artis Logicae*, 1849 ed., App. 64–66.

CHAPTER II

[1] "Introductory Lecture Delivered at the Opening of Classes in Mathematics at the University of London, 5 November, 1828," MS ADD. 3, University College Library, London, p. 12. For biographical information on De Morgan, see: Sophia Elizabeth De Morgan, *Memoir of Augustus De Morgan* (London: Longmans, Green, and Co., 1882); A.M.W. Stirling, *William De Morgan and His Wife* (New York: Henry Holt and Co., 1922); articles on De Morgan in the *Dictionary of National Biography*, *Dictionary of Scientific Biography*, and *Encylopaedia Brittanica*, 11th ed.; obituary notices in the *Athenaeum* (25 March 1871) and *Monthly Notices of the Royal Astronomical Society* (February 1872); Peter Heath's Introduction to *OS*; J.G. Crowther, *Scientific Types* (London: Barrie & Rockliff/The Cresset Press, 1969), 189–223; and Geoffrey Howson, *A History of Mathematics Education in England* (Cambridge: Cambridge University Press, 1982), 75–96.
[2] "Introductory Lecture," 17–18.
[3] De Morgan does not explicitly compare the meanings of "is" and "is equal to." We will see in Chapter 3 that he ascribes several meanings to "is." Equality in this geometrical context involves comparisons of magnitudes.
[4] The full title of this pamphlet was "First Notions of Logic, preparatory to the Study of Geometry." A few changes were made in reprinting it in FL, but they are not important for our purposes.
[5] Sophia De Morgan lists some 800 articles (*Memoir*, 407–14), which Leslie Stephen (*Dictionary of National Biography*) claims is roughly one-sixth of the *Cyclopaedia*. Stephen's fraction seems a bit high (there are some 14,000 double-columned pages in the *Cyclopaedia*), but De Morgan's contribution was still immense.
[6] *Penny Cyclopaedia*, Vol. XIX (1841), 372.
[7] *Loc. cit.*
[8] *Penny Cyclopaedia*, Vol. XXIII (1842), 437.
[9] *Op. cit.*, 438. The "sentences" which De Morgan gives are arguments in disguise rather than mere hypotheticals. An example is, "Every Y is X, every Z is Y; therefore every Z is X."

CHAPTER III

[1] For original publication information, see OS.
[2] De Morgan often referred to the fact that FL appeared on the same day as George Boole's *The Mathematical Analysis of Logic* (Cambridge: Macmillan, Barclay, & Macmillan, 1847). His first reference to this seems to be in his review of Boole, where he says, "On the very day in which Mr. De Morgan's work was published in London, the one at the head of our article appeared at Cambridge." (*Athenaeum*, 29 January 1848,

107.) This claim is essentially correct, though De Morgan may have stretched the truth by a day or two. He wrote in his own copy of Boole's work (University of London Library) that he received it on 27 November 1847. He wrote Boole on 28 November that he had read Boole's work "with great admiration," which suggests either that De Morgan read the work very quickly, or that he received it before the 27th, or that he was being overly polite. FL was published on 24 November. See G.C. Smith, *The Boole-De Morgan Correspondence 1842–1864* (Oxford: Clarendon Press, 1982), 24.

[3] De Morgan correctly adds the qualification that in the general case, we have "$y = -x$ as one at least of certain alternatives." Thus, suppose that $y = -x$ and $z = x^2$ yield $z = (-y)^2$. From this it follows only that $z = (-y)^2$ and $z = (x)^2$ yield *either* $y = -x$ or $y = x$. Thus, $y = -x$ is only one of two alternatives. But not even this condition is satisfied in the relational case (3.14). (A comment by John Corcoran pointed out to me the need for this clarification.)

[4] Unsigned review of FL, *Athenaeum* (11 December 1847), 1264. De Morgan does not identify the critic, but his phrasing is exactly the same as that of the review, except that there is a misprint which has replaced "marking" by "making."

[5] Once again, this is a reconstruction; De Morgan does not show how this is to be done.

[6] It must be confessed that De Morgan does not have a clearly stated distinction between premises and rules; but some such distinction seems required in this context. For an interesting discussion of this distinction in Boole, see John Corcoran and Susan Wood, "Boole's Criteria for Validity and Invalidity," *Notre Dame Journal of Formal Logic* 21 (1980): 609–38.

[7] C.S. Peirce, "Description of a Notation for the Logic of Relatives, Resulting from an Amplification of the Conceptions of Boole's Calculus of Logic," *Memoirs of the American Academy of Arts and Sciences*, n.s. Vol. 9, pt. 2 (1870): 317–78. On the relation between De Morgan and Peirce see my article, "De Morgan, Peirce and the Logic of Relations," *Transactions of the Charles S. Peirce Society* 14 (1978): 247–84.

[8] On earlier treatments of oblique inferences see Paul Thom, "Termini Obliqui and the Logic of Relations," *Archiv für Geschichte der Philosophie* 59 (1977): 143–55; E.J. Ashworth, "Joachim Jungius (1587–1657) and the Logic of Relations," *Archiv für Geschichte der Philosophie* 49 (1967): 72–85; and G.H.R. Parkinson, *Logic and Reality in Leibniz's Metaphysics* (Oxford: Oxford University Press, 1965), 37–50. Oblique inferences contain oblique terms–that is, terms in some grammatical case other than the nominative.

[9] De Morgan himself seems once to have claimed that oblique syllogisms could be treated syllogistically if one were given enough leeway in translation. In his article on "Syllogism" in the *Penny Cyclopaedia*, Vol. XXIII (1842), 439–40, he claimed that,

The thoughts of a man govern his actions.
John is a man.

John's thoughts govern his actions. ,

can be transformed into,

Every man is a being whose thoughts govern his actions.
John is a man.

John is a being whose thoughts govern his actions.

However, the application of this case to De Morgan's more well-known argument is unclear. Fifteen years after S2, in 1865, De Morgan was still issuing the challenge in the more public forum of the *Athenaeum* (May 13, June 3 and 24).

[10] Thus Alfred North Whitehead and Bertrand Russell appear to be incorrect in their treatment of the argument. *Principia Mathematica*, Vol. I, 2d ed. (1925; reprinted, Cambridge: Cambridge University Press, 1950), 291. For more on the formulation of this argument, see R.G. Wengert, "Schematizing De Morgan's Argument," *Notre Dame Journal of Formal Logic* 15 (1974): 165–66; and in response, my "On De Morgan's Argument," *Notre Dame Journal of Formal Logic* 18 (1977): 133–39.

[11] *Logic*, 50.

[12] S2, 28-29. Leibniz seems to have used a similar principle in dealing with oblique inferences. See Leibniz's 1687 letter to Vegetius, in G.H.R. Parkinson, ed., *Leibniz Logical Papers* (Oxford: Clarendon Press, 1966), 88.

[13] C.S. Peirce, "Description of a Notation for the Logic of Relatives," *passim*.

CHAPTER IV

[1] J.[oseph] S.[ortain], "De Morgan's Formal Logic," *Biblical Review* (1848): 406–07, 410–20; Sir William Hamilton, "On Affirmation and Negation–On Propositional Forms–On Breadth and Depth–On Syllogistic and Syllogistic Notation, &c.," in *Discussions on Philosophy and Literature*, 609–39; anon., *Athenaeum* (11 December 1847), 1263–65; and Henry L. Mansel, "Recent Extensions of Formal Logic," *North British Review* (1851): 90–121 (referred to hereafter and in the text by "M").

[2] The other two books were William Thomson, *An Outline of the Necessary Laws of Thought: A Treatise on Pure and Applied Logic* (London and Oxford, 1849); and Thomas Spencer Baynes, *An Essay on the New Analytic of Logical Forms* (Edinburgh and London, 1850).

[3] *New Analytic* (reprint, New York: Burt Franklin, 1971), 79.

[4] De Morgan states this point succinctly several years later, when he says, "A *form* is an empty machine, such as 'Every *man* is *animal*.' The logicians will not see that their *formal* proposition, 'Every X is Y,' is material in three points, the degree of assertion, the quantity of the proposition, and the copula. The purely formal proposition is 'There is the probability α that X stands in the relation L to Y.'" *A Budget of Paradoxes*, ed. Sophia De Morgan (London: Longmans, Green, and Co., 1872), 202–3.

[5] Formal properties of relations are those which can be defined solely in terms of variables and logical constants. For instance, R is *transitive* if and only if $(x)(y)(z)[(xRy \;\&\; yRz) \supset xRz]$. This definition requires the specification of the logical constants of a language, something which, as we will soon see, is an important issue for De Morgan.

[6] De Morgan continues to hold this position in S4 (217–18) and L (253–54).

[7] For this formulation see Corcoran and Wood, *op. cit.*, 612–13.

[8] Cf. the discussion of Reid in Chapter One.

[9] De Morgan's formulation here suggests an alternative interpretation of his treatment of this argument in FL. Here he refers not so much to quantified statements as to instances of them, and he may think of this as the locus of the substitution using his *dictum*. Rather than being, say, (3.44), the argument would be,

Every man is an animal.

Let X be the head of a man;
Then X is the head of an animal (by substitution).
Therefore, every head of a man is the head of an animal.

CHAPTER V

[1] As we shall see, Theorem K is the one exception to this.

[2] S4, 220–27. For discussions of this topic see: I.M. Bochenski, *A History of Formal Logic*, tr. Ivo Thomas (Notre Dame, Indiana: University of Notre Dame Press, 1961), 375–77; Jorgen Jorgensen, *A Treatise of Formal Logic* Vol. I (Copenhagen: Levin & Munksgaard, 1931), I, 92–97; William Kneale and Martha Kneale, *The Development of Logic* (Oxford: Clarendon Press,1962), 427–28; C.I. Lewis, *A Survey of Symbolic Logic* (1918; reprint, New York: Dover Publications, 1960), 43–51; A.N. Prior, *Formal Logic* (Oxford: Clarendon Press, 1955), 152–54; Boruch A. Brody, "The Rise of the Algebra of Logic" (diss., Princeton University, 1967), 184–202; Carl Gottschall, "An Examination and Evaluation of the Historical Importance of the Principal Innovative Contributions to Formal Logic of Augustus De Morgan" (diss., New York University, 1980), 149–70; Benjamin S. Hawkins, "A Reassessment of Augustus De Morgan's Logic of Relations: A Documentary Reconstruction," *International Logic Review* 10 (1979): 32–61; R.M. Martin, "De Morgan and the Logic of Relations," in *Peirce's Logic of Relations and Other Studies* (Lisse: Peter de Ridder Press, 1979), 46–53.

[3] As stated by De Morgan, this may carry the presupposition that there is at least one L of Y. This is not intended. A better version might be "X is not an L of Y."

[4] See "On the Foundation of Algebra, No. IV., on Triple Algebra," *Transactions of the Cambridge Philosophical Society* 8 (1842–47): 241.

[5] Alfred Tarski, "On the Calculus of Relations," *Journal of Symbolic Logic* 6 (1941): 73–89; for the more recent approach see Alfred Tarski and Steven Givant, *A Formalization of Set Theory Without Variables* (Providence: American Mathematical Society, 1987), esp. 45–51. Aristotle also uses individual cases to establish the basic general laws of his logic.

[6] De Morgan proposed this as a puzzle in *Notes and Queries* (14 January 1860): 25, and reported on its solution in *Notes and Queries* (10 March 1860): 184–85. He also posed the problem to Sir William Rowan Hamilton. See De Morgan to Hamilton, 22 August 1859; Hamilton to De Morgan, 26 August 1859 (misdated 1850); and De Morgan to Hamilton, 7 September 1859 (Sir William Rowan Hamilton Correspondence, Library of Trinity College, Dublin). He also sent the problem to Sir John Herschel (Herschel Correspondence, Royal Society of London). This letter is undated, but De Morgan's address on it means that it was not written before early summer of 1859.

[7] I owe this point to John Corcoran (private communication), who suggests that De Morgan's conjecture might be modified to read, that every relation which is convertible and weakly reflexive satisfies condition D. I do not know whether this conjecture is true or false.

[8] Bertrand Russell claims that De Morgan was the first person to use "transitive" in this sense. See *Principles of Mathematics*, 2nd ed. (New York: W.W. Norton & Co., 1938), 218n.

[9] "Logic Notebook," Ms. 399, 50r; in the C.S. Peirce Papers, Harvard University.

[10] "Description of a Notation for the Logic of Relatives."

CHAPTER VI

[1] Omitted accents have been added.
[2] The basic idea of this goes back to FL, 90–91 and 334–38.
[3] Paul Lorenzen, *Formal Logic*, tr. F. Crosson (Dordrecht: D. Reidel Publishing Co., 1965), 1–18. See also Günther Patzig, *Aristotle's Theory of the Syllogism*, tr. J. Barnes (Dordrecht: D. Reidel Publishing Co., 1968), 49–57. The Gergonne Relations can also be looked at in this way. See Kneale and Kneale, *The Development of Logic*, 350–52; and J.A. Faris, "The Gergonne Relations," *Journal of Symbolic Logic* 20 (1955): 207–31.

CHAPTER VII

[1] *Encylopedia Brittanica*, 11th ed., *s.v.* "De Morgan, Augustus."
[2] See his wide-ranging contributions to the *Penny Cyclopaedia*, *Notes and Queries*, and, especially, the *Athenaeum*. Even though his contributions to the *Athenaeum* were unsigned, it is possible to determine what they were because bound copies of them, initialed by De Morgan, are deposited at the University of London Library.
[3] "On the Foundation of Algebra," *Transactions of the Cambridge Philosophical Society* 7 (1841): 173–87; 7 (1842): 287–300; 8 (1844): 139–42; 8 (1847): 241–53.
[4] In fact, only one of the 324 teaching notebooks deals solely with logic. This is Ms. 775/203, "On the Two Forms of the Universal Affirmative Proposition," De Morgan Papers, University of London Library. A few other notebooks consider geometrical demonstration (see note 23).
[5] "On the Study of Mathematics."
[6] De Morgan even uses this very language. See S5, 345 and De Morgan's review of Herbert Spencer, *The Classification of the Sciences*, *Athenaeum* (21 May 1864): 709.
[7] See note 1, Chapter Two.
[8] See also De Morgan's review of Sir William Hamilton, *Lectures on Metaphysics and Logic*, *Athenaeum* (24 November 1860): 706.
[9] See also *Notes and Queries* (27 August 1864): 161.
[10] See also De Morgan's review of John Stuart Mill, *An Examination of Sir William Hamilton's Philosophy*, *Athenaeum* (27 May 1865): 711; "Speech of Professor De Morgan, President," *Proceedings of the London Mathematical Society* 1 (1866): 1–9.
[11] *Penny Cyclopaedia*, Vol. XV (1839), *s.v.* "Mathematics," 11ff.
[12] G.C. Smith, ed., *The Boole-De Morgan Correspondence 1842-1864* (Oxford: Clarendon Press, 1982).
[13] For some other references to Boole by De Morgan, see his review of Boole's *Mathematical Analysis of Logic*, *Athenaeum* (29 January 1848): 107–8; his obituary of Boole, *Macmillan's Magazine* (February, 1865): 279–80; "Budget of Paradoxes, No. XXIII," *Athenaeum* (11 March 1865): 350; De Morgan to Sir William Rowan Hamilton, 5 October 1852, in Robert Percival Graves, *Life of Sir William Rowan Hamilton*, Vol. III (Dublin: Hodges, Figgis & Co./London: Longmans, Green & Co., 1889), 421.
[14] In the title of his detailed review of *The Boole-De Morgan Correspondence*, John Corcoran has gone so far as to use the apt phrase "Correspondence without Communication"; *History and Philosphy of Logic* 7 (1986): 65–75. For other reviews see those by Theodore Hailperin, *Journal of Symbolic Logic* 49 (1984): 657–59; and Calvin Jongsma,

Historia Mathematica 12 (1985): 186–90. In reading through De Morgan's extensive correspondence with both Boole and Sir William Rowan Hamilton, one gets the impression that he never seriously studied their works. With his great variety of interests and heavy work schedule, he probably had neither the time nor the inclination to do so. Boole seems not to have studied De Morgan's work carefully either. On 27 May 1852, he wrote Sir William Hamilton, "The peculiarities of Mr. De Morgan's system of logic I have not made an object of study, and I do not feel competent therefore to pass any opinion upon the correctness or mutual consistency of his views." John Veitch, *Memoir of Sir William Hamilton, Bart.* (Edinburgh and London: William Blackwood and Sons, 1869), 344.

[15] De Morgan to Boole, 28 November 1847, *Correspondence*, 25.

[16] "Introductory Lecture" and *Penny Cyclopaedia*, Vol. XI (1838), *s.v.* "Geometry," 151, 156.

[17] In this we followed Ian Mueller, "Greek Mathematics," 37–43.

[18] These include *The Connexion of Number and Magnitude: An Attempt to Explain the Fifth Book of Euclid* (London: Taylor and Walton, 1836); "Short Supplementary Remarks on the First Six Books of Euclid's Elements," *Companion to the Almanac for* 1849, 5–20; "On Indirect Demonstration," *Philosophical Magazine* 3 (1852): 435–38. Among his teaching notebooks in the University of London Library, see "Notions Preliminary to Geometry," 775/202; "On the Two Forms of the Universal Affirmative Proposition," 775/203; and 775/310 (untitled).

[19] For De Morgan's views on issues regarding the place of logic in the teaching of geometry, see the following articles from the *Quarterly Journal of Education*: "On Mathematical Instruction" 1 (1831): 264– 79; "On the Method of Teaching the Elements of Geometry, Part I" 6 (1833): 35–49; and "A Method of Teaching Geometry, No. II" 6 (1833): 237–51.

[20] *Penny Cyclopaedia*, Vol. XV (1839), *s.v.* "Mathematics," 12.

[21] See especially the "Foundation of Algebra" series (n. 3).

[22] "Calculus of Functions," *Encyclopaedia Metropolitana* (1836), 306–92; and *Penny Cyclopaedia*, Vol. XVI (1840), *s.v.* "Operation."

[23] For the problems of interpreting negative and "impossible" numbers see Ernest Nagel, " 'Impossible Numbers': A Chapter in the History of Modern Logic," in *Studies in the History of Ideas*, ed. Columbia University Philosophy Department (New York: Columbia University Press, 1935), 429–74; Morris Kline, *Mathematical Thought from Ancient to Modern Times* (New York: Oxford University Press, 1972), 592–97, 626–32.

[24] *The Principles of Algebra*, 2 vols. (London, 1796–99).

[25] We will leave open the question of just which laws to retain. We must, for instance, exclude laws such as $(x) \sim (x^2 = -1)$, which hold for the reals and not for the imaginaries. Peacock seems to have had in mind algebraic identities, such as $x(x + 1) = x^2 + x$.

[26] For recent discussions of this period see Morris Kline, *Mathematical Thought*, 772–94; Harvey W. Becher, "Woodhouse, Babbage, Peacock and Modern Algebra," *Historia Mathematica* 7 (1980): 389–400; Daniel A. Clock, "A New British Concept of Algebra: 1825–1850" (diss., University of Wisconsin, 1964); J.M. Dubbey, "Babbage, Peacock and Modern Algebra," *Historia Mathematica* 4 (1977): 295–302; Elaine Koppelman, "The Calculus of Operations and the Rise of Abstract Algebra," *Archive for History of*

Exact Sciences 8 (1971): 155–242; Luboš Nový, *Origins of Modern Algebra* (Leyden: Noordhoff International Publishing, 1973); Helena M. Pycior, "George Peacock and the British Origins of Symbolical Algebra," *Historia Mathematica* 8 (1981): 23–45; "Historical Roots of Confusion among Beginning Algebra Students: A Newly Discovered Manuscript," *Mathematics Magazine* 55 (1982): 150–56; "Early Criticism of the Symbolical Approach to Algebra," *Historia Mathematica* 9 (1982): 392- 412; "Internalism, Externalism, and Beyond: 19th Century British Algebra," *Historia Mathematica* 11 (1984): 424–41; Joan L. Richards, "The Art and Science of British Algebra: A Study in the Perception of Mathematical Truth," *Historia Mathematica* 7 (1980): 343–65.

[27] *A Treatise on Algebra*: Vol. I., *Arithmetical Algebra* (Cambridge: Cambridge University Press, 1842); Vol. II., *On Symbolical Algebra and its Applications to the Geometry of Position* (Cambridge: Cambridge University Press, 1845); "Report on the Recent Progress and Present State of Certain Branches of Analysis," *Report of the Third Meeting of the British Association for the Advancement of Science* (London: John Murray, 1834), 185–352. See also De Morgan's review of Peacock's *Treatise* in the *Quarterly Journal of Education* 9 (1835): 91- 110, 293–311.

[28] "Report," 194.

[29] "Report," 194–95.

[30] "Report," 198.

[31] "Report," 199.

[32] "Report," 200.

[33] *Trigonometry and Double Algebra* (London: Taylor, Walton, and Maberly, 1849), 89. This will be refered to as "TDA." For discussions of De Morgan's work in algebra see Helena M. Pycior, "Augustus De Morgan's Algebraic Work: The Three Stages," *Isis* 74 (1983): 211–26; Joan L. Richards, "Augustus De Morgan, The History of Mathematics, and the Foundations of Algebra," *Isis* 78 (1987): 7- 30; and G.C. Smith, "De Morgan and the Laws of Algebra," *Centaurus* 25 (1981): 50–70.

[34] "On Infinity; and on the Sign of Equality," *Transactions of the Cambridge Philosophical Society* 11 (1866): 145- 89; "On Divergent Series and Various Points of Analysis Connected with Them," *Transactions of the Cambridge Philosophical Society* 8 (1849): 182- 203.

CHAPTER VIII

[1] "Calculus of Relations."

[2] *Loc. cit.* For an early formulation of the logic of relations in this style, see L.H. Chin and A. Tarski, "Distributive and Modular Laws in the Arithmetic of Relation Algebras," *University of California Publications in Mathematics*, new series 1 (1951): 341- 84. See also B. Jónsson, "Varieties of Relation Algebras," *Algebra Universalis* 15 (1982), 273–98.

[3] T. Smiley, "What is a Syllogism?", *Journal of Philosophical Logic* 2 (1973): 136–54; John Corcoran, "Aristotle's Natural Deduction System," in J. Corcoran (ed.), *Ancient Logic and Its Modern Interpretations* (Dordrecht: D. Reidel Publishing Co., 1973), 85–131.

[4] The basic idea for this derivation comes from Chin and Tarski, *op. cit.*, 355.

[5] See for instance, A.Tarski and S. Givant, *loc. cit.*

[6] *Ibid.*, 50.
[7] *Ibid.*, 50.
[8] The derivations of DR10.8 and 10.9 are adapted from Chin and Tarski, *op. cit.*, 357. Both use P2 essentially. Unfortunately, their proof of P2 makes substantial use of the Boolean part of the full logic of relations, which we do not have available.
[9] Cf. J. Corcoran, *op. cit.*, 99.
[10] See T. Smiley, *op. cit.* and J. Corcoran, *op. cit.*
[11] John Corcoran called my attention to the importance of De Morgan's 1864 annotation, and he pointed out the relationship between De Morgan's second conjecture and the Axiom of Choice.

INDEX

Abstraction and generalization, 100
Agreement, in particulars, 52, 73
Aldrich, Henry, 2, 29
Algebra, 100, 170; arithmetic, 187–88; double, 189–92; form and matter in, 184; and logic, 60, 63–64, 71–72, 85, 182, 192; nature of, 180; 19th century British, 184, 186; reasoning in, 170, 180–81, 194; rigor in, 170–71, 180; single, 189– 90; symbolic, 187–90; use in logic, 174, 190. *See also* Arithmetic, Mathematics
Alternative, 169
Alternatives, contradictory, 52
Argand, Jean Robert, 185
Arguments: a fortiori, 19, 26, 33–35, 39–43, 47, 59–60; inductive, 10; linguistic form of, 4; nature of, 4; reduction to syllogistic form, 26; validity of, 5, 8. *See also* Inferences, Reasoning
Aristotle, 79, 100, 115, 124, 176, 200, 229, 232
Arithmetic, 117; particular and universal, 189
Assertion and denial, 36, 38, 117, 135
Associativity, 119, 238–39
Assumptions, existential, 139
Attributes, 151
Axiom of Choice, 243–44
Axiom 6, 209, 219, 238

Baynes, Thomas Spencer, 100
Boole, George, 3, 170–71, 174–75, 232, 246 n14, 247 n2, 252 n14
Brother, reflexivity of, 127
Buée, Abbé Adrien-Quentin, 185

Calculus of Functions, 182, 194
Calculus, symbolic, 188–90
Chin, L.H., 253 n2
Class, 151; null and universal, 229; partition of, 243
Co-inadequate, 166
Complement, 57, 166
Composition: of functions, 182, 194; of relations, 42, 49, 58, 68–75, 78, 89, 97, 104–5, 113–14, 117, 132, 134, 140, 147, 150, 169, 181–83, 194, 199; of relations and classes, 114, 138, 140, 142
Conceiving, formal, 92
Conclusion, negative, 132, 144
Consequence, formal, 111–12
Contraposition, 114, 146–47, 178
Contrary: of relations, 113, 117, 132, 199; of terms, 77, 91, 152–54
Conventions, linguistic, 127–28
Converse, of relations, 113, 117, 132, 143, 150, 158, 164, 183, 199
Conversion of relations, 16, 21, 25
Convertibility: of copulas, 64–65, 67, 69, 73, 100, 150, 157–58, 161, 169; of identity, 95, 150, 155–56; of *is*, 52, 57, 95, 116; of relations, 124–25, 144, 177; and transitivity, 238. *See also* Relations
Copulas, 151; abstract, 48, 50, 54, 56, 63–64, 69, 73, 78, 89, 90- -91, 94, 104–5, 150, 157, 169, 179–80, 194; contrariety of, 64; correlative (converse), 67, 72, 77; essential characteristics of, 64; form and matter in, 98–100, 102; formal properties of, 64–65, 100, 102, 108, 116, 158, 160, 179; inconvertible, 160, 164, 169; multiple, 26, 28–29, 48, 54, 94; relational analysis of, 30, 35, 107–8; two, 29, 48, 55, 179. *See also* Equality; *is*; Relations
Corcoran, John, 200
Cross-operator, 237, 239
Counterexamples, method of, 5, 7

Dasypodius, Cunradus, 13–15, 17
Daughter, reflexivity of, 129
De Morgan, Augustus, 26–27, 43, 48, 50, 114–15, 143, 170–71, 174–76, 180, 185, 196, 221, 244, 247 n1, 247 n3, 247 n2, 250 n8, 251 n2, 252 n14
Decomposition of relations, 133, 135
Deficient, 166
Dependent, 169
Dictum: de majore et minore, 67–69, 71, 80–81, 110; *de omni et nullo*, 5, 10, 49, 80–88, 105–6; relational, 106
Dot-operator, 236 Duncan, William, 2

Elimination and inference, 61–62, 72, 85, 175, 182
Enunciation, modes of, 165, 169
Equality, 44; as a copula, 26, 28–29; as predicate and relation, 30–31; reduction to identity, 55
Equivalence, relational, 118, 123, 199, 202
Essential, 169
Euclid, 176, 178. *See also* Geometry
Excluded middle, law of, 116
Exient, 166
Existence, assertion of, 54, 235
Existence Condition, 124, 210–12, 225–27, 234
Exponents, fractional, 192
External, 57, 166

Figures, 156–57, 160, 162; first, 130–132, 157–58; fourth, 72, 145; second, 131
Force, of expression, 4, 8
Form: as abstraction and generalization, 1, 89, 99, 170, 192, 194; logical, 1, 3–10, 15, 116, 194; and matter, 89, 91–92, 95, 99, 117; and matter, in logic, 184, 192; and matter, in mathematics, 100, 184, 193; sameness of, 5, 104, 106
Forms: opponent, 65; propositional, 1, 9, 37, 46–47, 57, 63, 138, 153, 157, 160–61, 165, 167; quantificational and singular, 66; syllogistic, 150, 167 Frege, Gottlob, 107, 109, 171
Frend, William, 185, 187
Functionality Condition, 226, 229
Functions, 23

Gauss, Karl Friedrich, 185
Geometry: nature of, 13, 17, 27, 173; reasoning in, 12, 28, 170, 176–80, 194; use in logic, 174, 194; use of logic in, 2,10–15, 24, 28, 32, 171, 176–78, 194. *See also* Mathematics
Generalization, 71; rule of, 137
Genus, 57, 166
Givant, Steven, 200
Greater than, analysis of, 34
Grouping, 119

Hamilton, Sir William, 2, 13, 15–22, 25, 29–30, 32, 63, 77, 90, 96–97, 100, 172, 179
Hamilton, Sir William Rowan, 188, 252 n14
Hartley, David, 130
Herlinus, Christian, 13–15, 17

Identity, 36–38, 44, 50, 52, 54, 73–76, 93–94, 98, 108, 117, 155; law of, 93, 116; logic of, 196, 221–23; reduction to, 54, 74- -75. *See also* Copula, *is*
Import, existential, 78, 111, 124, 129
Inalternative, 169 Inclusion, relational, 118, 199
Independent, 169
Inessential, 169
Inferences: elliptical, 97; forms of, 103; identity theory of, 50; immediate, 166, 178; material, 97, 103; oblique, 32, 49, 59, 79–91, 110, 114, 146–47, 248 n9, 249 n12; relational, 1, 16; relational theory of, 39, 43–44, 50, 195. *See also* Arguments, Reasoning
Innovations, in logic, 26, 63, 84, 87, 91, 103, 118

Instantiation, rule of, 137
Intention, first and second, 150–51
Interpretation: mathematical subjective, 150, 166; metaphysical subjective, 168; of symbols in mathematics, 189–92
Inverses, 123, 248 n3
Irrepugnant, 169
is: as copula, 28–29; formal laws of, 51, 56; functions of, 109, 117; materiality of, 98; senses of, 51–54, 93–94. *See also* Copula

Judging, formal, 92
Judgments, 152; categorical, 93; pluralistic theory of, 107; purely formal, 103; relational analysis of, 94, 106–8; subject-predicate analysis of, 90, 94, 106–7. *See also* Propositions
Jungius, Joachim, 79

Kant, Immanuel, 100

Laws: commutative, 188; De Morgan's, 146; formal, 193; retention of, 252 n25; of thought, necessary, 173
Leibniz, Gottfried Wilhelm, 13, 15, 249 n12
Locke, John, 2
Logic: independence of geometry, 11, 176; and language, 9; mathematical, 170–72; and mathematics, 11, 91, 170–73, 177; and mathematics, analogies between, 175, 194; nature of, 1–4, 7–9, 91, 93, 100, 103–4; of relations, construction of, 197; of relations, and mathematics, 194; of relations, nature of, 124, 195; of relations, and traditional syllogism, 114, 117, 124; sentential, 200
Lorenzen, Paul, 167
Lukasiewicz, Jan, 200

Mansel, Rev. Henry L., 2, 19–25, 29, 40, 89–97, 100–1, 104–8, 111, 179
Maseres, Baron Francis, 187
Mathematics: conceptual problems of, 27; and education, 27, 171–72; nature of, 172, 176, 184; proof and logic in, 16, 25, 147. *See also* Algebra, Arithmetic, Geometry
Matter, of thought, necessary, 173
Mueller, Ian, 11, 15, 17

Negation, absence of, 231–33 Non-contradiction, law of, 116
Non-inclusion and non-identity, 233
Notation, 117, 140–42, 157; spicular, 63, 78, 155, 165
Numbers: complex, 185–86; imaginary (impossible), 170, 184–85, 189–92; negative, 170–71, 184–85

Objects, 150
Ockham, William of, 115
Opposition, formal laws of, 60–61

Partient, 166
Peacock, George, 186–88, 190
Peirce, Charles S., 86, 107, 143, 197, 221
Phases: first, 144–45; primary, 134, 145, 162–64; syllogistic, 162
Precision, terminal, 156
Predication, 75, 83
Premises, affirmative and negative, 131, 135
Principium et exemplum, 177
Principle, of the Permanence of Equivalent Forms, 188, 190
Principles: formal and material, 20, 22–24, 30, 40, 97–98; logical, as premises, 18, 20, 40, 68–69, 97–98, 106, 248 n6; role in geometry, 17–22
Probability, theory of, 90–91, 101, 175
Product, relative, 167
Properties, formal, 192, 249 n4
Propositions: affirmative and negative, 36, 152, 154; analysis of, 38–39, 57; categorical, 16, 50, 93, 109, 232–33; categorical,

reduction to, 36; hypothetical, 38; identity, analysis of, 37; purely formal, 101, 103, 249 n4; relational analysis of, 32, 35, 57, 101, 106–10, 152, 164, 195; relational, analysis of 2, 16, 101, 106, 109; relational and categorical, 16; subject-predicate analysis of, 29; transformations of, 114, 145–47, 153, 177; unity of, 109–10. *See also* Judgments
Pythagorean Theorem, 28, 30–32

Qualities, 150; possession of, 53
Quantification: instances of, 249 n9; of subject and predicate, 77, 90; theory of, 196–97, 229–35
Quantifiers, 21, 102, 118–19
Quantity, inherent, 113, 117, 132, 134, 194, 199
Quaternions, 188

Readings, cumular and exemplar, 152, 156
Reasoning: and comparison, 45, 47, 49; everyday forms of, 145; formal, 92; in mathematics, 124; nature of, 4, 94; purely formal, 117; relational theory of, 35. *See also* Arguments, Inferences
Reductio ad absurdam, 132, 135, 232, 237
Reid, Thomas, 2, 15–17, 21, 24–25, 29–30, 107, 178–79
Relations, 151; asymmetric, 34; and comparison, 35–36, 39; convertible, 113; convertible, reducibility of, 125–26, 240; convertible and reflexive, 125; convertible and transitive, 129; convertible and transitive, reduction of, 235, 240–44; convertible and weakly reflexive, reduction of, 240–41, 250 n7; equivalence, 126, 224, 243; formal and material, 108; formal properties of, 192, 249 n4; formulation of logic of, 119; general logic of, 76, 113, 115, 117–24; in logic and mathematics, 183; null, 200, 230–32; onymatic, 108–9, 115, 117, 146–47, 152, 158–59, 165; primitive, 167; quantified, 131–32, 136; reduction to identity, 44–46; reflexive, 113, 124, 126–27; between subject and predicate, 108; three-place, 110; transitive, 113, 128; universal, 230–32; weakly reflexive, 126. *See also* Convertibility, Copula, *is*, Transitivity
Replacement, rule of, 122
Repugnant, 169
Resemblance, 44, 46
Resolution, of relations, 72–73
Roots: complex, 191; imaginary, 185–86; negative, 186

Sameness, 44
Schema, method of abstract, 6–7
Schröder, Ernst, 197, 221
Science: demonstrative, 11, 24, 176, 180; exact, 173; formal, 187
Signs, formal, 98
Sortain, Joseph, 90 Smiley, T., 200
Species, 53, 57, 96–97, 166
Substitution, rule of, 79–80, 83–88
Statements, of fact and taste, 76
Subjects, complex logical, 17, 24
Subsumption, multiple, 21–22, 24–25, 31, 40
Syllogism: arithmetic form of cumular, 155; bicopular, 49, 69–70, 73, 78, 104, 114, 150, 194; complex, 59–60; and copular laws, 65; demonstrative, 24; formal theory of, 194; forms of, 154–55; hypothetical, 10–11, 37; limited unit, 114, 143–44; nature of, 4, 46, 109, 130; numerically definite, 59, 91, 101, 175; onymatic, 147, 194; opponent, 72–73; ordinary, 145; quantified relational, 136–37, 142–43; regular, 166; relational, 130, 145, 147, 156, 159, 194; relational, and algebra, 182–83; relational, and middleterm,

146–47; strengthened, 166; traditional, 84, 109, 146,194; traditional, adequacy of, 2–3, 10–11, 16–17, 23, 25–26, 37, 43, 46–47, 49, 58–59, 116, 176–77; traditional, reduction to, 31, 33–34, 39–43, 52–53, 59, 67, 70, 79, 96; traditional, reduction to identity, 56; traditional, relational analysis of, 1, 37, 56, 106, 143, 148–49, 152, 166, 194; unicopular, 72–73; unit, 129–36, 143, 145, 155–56; unit, figures and phases of, 130–35; unit, forms of, 134; unit, reduction to, 137–38; validity of, 98–99, 105, 109
Symbols: Boolean, 197–98, 237–38; Peircean, 197–99
System: Deductive, 196; D, abbreviation of proofs, 205; D, added principles of, 227–28, 230–31; D, basic identities of, 206– 8; D, convertible relations, 215–16; D, convertible and transitive relations, 215–17; D, formulation of 196–200; D, inclusion and identity laws, 202–3, 217–18; D, primitive basis, 200–2; D, reflexive relations, 217; D, rule of replacement, 204–5; D, Theorem K, 208–10, 238; D, transitive relations, 212–17; DI (with identity), 221–29; DI, quantification theory in, 229–35; DI, reflexivity in, 222; DI, with uniqueness operator, 238–40; DIN, added principles of, 234; DIN, quantification theory in, 232–35

Tarski, Alfred, 119, 197–200, 221
Terms, 151; combinations of, 215–17, 219–21, 223–24; complex, 114, 145–47; complex, substitution in, 84–87; particular and universal use of, 82, 86–88; three-relation, 219–21; total and partial, 152
Theorem K, 113, 120–22, 150, 166, 169, 208–10, 238
Thinking, formal, 93
Thought, forms of, 62, 100
Transitivity: of copula, 64–65, 68–69, 73, 99–100, 102, 150, 157–58, 169, 193; of equality, 2, 15–16, 20, 29–32, 48–49, 96–98, 101, 178–79, 194; of *greater than*, 19–20, 33–34, 55–56, 60, 179; of identity, 105, 150, 155–56, of *is*, 52– 53, 56, 69, 101, 104–5, 116; of *less than*, 39–41, 43, 55; of relational inclusion, 106; of relations, 2, 11, 23, 25, 32–33, 47, 56, 67, 89, 94, 104–5, 143–44, 161, 179; of species relation, 96, 98, 105. *See also* Copula, *is*, Relations
Trigonometry, 192
Triples, number, 171

Uniqueness: clauses, 242–45; operator, 196, 235–40

Validity: of arguments, 8; formal, 97, 104, 110–11, 192; formal and material, 89–90, 95, 111; intuitions of, 87
Variables, use of, 6,21
View: objective, 149–51, 165; subjective, 149–52, 164–69

Watts, Isaac, 2
Weak reflexivity, 240–41, 243
Whately, Archbishop Richard, 1–10, 20, 93, 106
Words, logical and descriptive, 9, 106–7

Synthese Historical Library

Texts and Studies in the History of Logic and Philosophy

Series Editor: Norman Kretzmann (*Cornell University*)

1. M.T. Beonio-Brocchieri Fumagalli: *The Logic of Abelard.* Translated from Italian by S. Pleasance. 1969 ISBN 90–277–0068–0
2. G. W. Leibniz: *Philosophical Papers and Letters.* A Selection, translated and edited, with an Introduction, by L. E. Loemker. 2nd ed., 2nd printing. 1976 ISBN 90–277–0008–8
3. E. Mally: *Logische Schriften.* Grosses Logikfragment – Grundgesetze des Sollens. Herausgegeben von K. Wolf und P. Weingartner. 1971 ISBN 90–277–0174–1
4. L. W. Beck (ed.): *Proceedings of the Third International Kant Congress.* 1972 ISBN 90–277–0188–1
5. B. Bolzano: *Theory of Science.* A Selection with an Introduction by J. Berg. Translated from German by B. Terrell. 1973 ISBN 90–277–0248–9
6. J. M. E. Moravcsik (ed.): *Patterns in Plato's Thought.* 1973 ISBN 90–277–0286–1
7. Avicenna: *The Propositional Logic.* A Translation from *Al-Shifā': al-Qiyās*, with Introduction, Commentary and Glossary by N. Shehaby. 1973 ISBN 90–277–0360–4
8. D. P. Henry: *Commentary on* De Grammatico. *The Historical-Logical Dimensions of a Dialogue of St. Anselms's.* 1974 ISBN 90–277–0382–5
9. J. Corcoran (ed.): *Ancient Logic and its Modern Interpretations.* 1974 ISBN 90–277–0395–7
10. E. M. Barth: *The Logic of the Articles in Traditional Philosophy.* A Contribution to the Study of Conceptual Structures. 1974 ISBN 90–277–0350–7
11. J. Hintikka: *Knowledge and the Known.* Historical Perspectives in Epistemology. 1974 ISBN 90–277–0455–4
12. E. J. Ashworth: *Language and Logic in the Post-Medieval Period.* 1974 ISBN 90–277–0464–3
13. Aristotle: *The Nicomachean Ethics.* Translation with Commentaries and Glossary by H. G. Apostle. 1974 ISBN 90–277–0569–0
14. R. M. Dancy: *Sense and Contradiction.* A Study in Aristotle. 1975 ISBN 90–277–0565–8
15. W. R. Knorr: *The Evolution of the Euclidean Elements.* A Study of the Theory of Incommensurable Magnitudes and its Significance for Early Greek Geometry. 1975 ISBN 90–277–0509–7
16. Augustine: *De Dialectica.* Translated with Introduction and Notes by B. D. Jackson from the Text newly edited by J. Pinborg. 1975 ISBN 90–277–0538–9
17. Á. Szabó: *The Beginnings of Greek Mathematics.* Translated from German. 1978 ISBN 90–277–0819–3
18. Juan Luis Vives: *Against the Pseudodialecticians.* A Humanist Attack on Medieval Logic. Texts (in Latin), with Translation, Introduction and Notes by R. Guerlac. 1979 ISBN 90–277–0900–9
19. Peter of Ailly: *Concepts and Insolubles.* An Annotated Translation (from Latin) by P. V. Spade. 1980 ISBN 90–277–1079–1
20. S. Knuuttila (ed.): *Reforging the Great Chain of Being.* Studies of the History of Modal Theories. 1981 ISBN 90–277–1125–9
21. J. V. Buroker: *Space and Incongruence.* The Origin of Kant's Idealism. 1981 ISBN 90–277–1203–4

Synthese Historical Library

22. Marsilius of Inghen: *Treatises on the Properties of Terms*. A First Critical Edition of the *Suppositiones, Ampliationes, Appellationes, Restrictiones* and *Alienationes* with Introduction, Translation, Notes and Appendices by E. P. Bos. 1983 ISBN 90–277–1343–X
23. W. R. de Jong: *The Semantics of John Stuart Mill*. 1982 ISBN 90–277–1408–8
24. René Descartes: *Principles of Philosophy*. Translation with Explanatory Notes by V. R. Miller and R. P. Miller. 1983 ISBN 90–277–1451–7
25. T. Rudavsky (ed.): *Divine Omniscience and Omnipotence in Medieval Philosophy*. Islamic, Jewish and Christian Perspectives. 1985 ISBN 90–277–1750–8
26. William Heytesbury: *On Maxima and Minima*. Chapter V of *Rules for Solving Sophismata*, with an Anonymous 14th-century Discussion. Translation from Latin with an Introduction and Study by J. Longeway. 1984 ISBN 90–277–1868–7
27. Jean Buridan's *Logic. The Treatise on Supposition. The Treatise on Consequences*. Translation from Latin with a Philosophical Introduction by P. King. 1985 ISBN 90–277–1918–7
28. S. Knuuttila and J. Hintikka (eds.): *The Logic of Being*. Historical Studies. 1986 ISBN 90–277–2019–3
29. E. Sosa (ed.): *Essays on the Philosophy of George Berkeley*. 1987 ISBN 90–277–2405–9
30. B. Brundell: *Pierre Gassendi: From Aristotelianism to a New Natural Philosophy*. 1987 ISBN 90–277–2428–8
31. Adam de Wodeham: *Tractatus de indivisibilibus*. A Critical Edition with Introduction, Translation, and Textual Notes by R. Wood. 1988 ISBN 90–277–2424–5
32. N. Kretzmann (ed.): *Meaning and Inference in Medieval Philosophy*. Studies in Memory of J. Pinborg (1937–1982). 1988 ISBN 90–277–2577–2
33. S. Knuuttila (ed.): *Modern Modalities*. Studies of the History of Modal Theories from Medieval Nominalism to Logical Positivism. 1988 ISBN 90–277–2678–7
34. G. F. Scarre: *Logic and Reality in the Philosophy of John Stuart Mill*. 1988 ISBN 90–277–2739–2
35. J. van Rijen: *Aspects of Aristotle's Logic of Modalities*. 1989 ISBN 0–7923–0048–3
36. L. Baudry: *The Quarrel over Future Contingents (Louvain 1465–1475)*. Unpublished Latin Texts collected and translated in French by L. Baudry. Translated from French by R. Guerlac. 1989 ISBN 0–7923–0454–3

THE NEW SYNTHESE HISTORICAL LIBRARY
Texts and Studies in the History of Philosophy

37. S. Payne: *John of the Cross and the Cognitive Value of Mysticism*. An Analysis of Sanjuanist Teaching and its Philosophical Implications for Contemporary Discussions of Mystical Experience. 1990 ISBN 0-7923-0707-0
38. D.D. Merrill: *Augustus De Morgan and the Logic of Relations*. 1990 ISBN 0-7923-0758-5

Kluwer Academic Publishers – Dordrecht / Boston / London

GPSR Compliance
The European Union's (EU) General Product Safety Regulation (GPSR) is a set of rules that requires consumer products to be safe and our obligations to ensure this.

If you have any concerns about our products, you can contact us on

ProductSafety@springernature.com

In case Publisher is established outside the EU, the EU authorized representative is:

Springer Nature Customer Service Center GmbH
Europaplatz 3
69115 Heidelberg, Germany

www.ingramcontent.com/pod-product-compliance
Lightning Source LLC
LaVergne TN
LVHW011752200926
848484LV00003B/685

* 9 7 8 0 7 9 2 3 0 7 5 8 7 *